Envisioning the Future of Mining

Envisioning the Future of Mining

Editors

Juan M Menéndez Aguado
Oscar Jaime Restrepo Baena
Jessica M. Smith

MDPI

Basel • Beijing • Wuhan • Barcelona • Belgrade • Novi Sad • Cluj • Manchester

Editors
Juan M Menéndez Aguado
Escuela Politécnica de Mieres
Universidad de Oviedo
Mieres
Spain

Oscar Jaime Restrepo Baena
Department of Materials and Minerals
Universidad Nacional de Colombia
Medellin
Colombia

Jessica M. Smith
Department of Engineering, Design & Society
Colorado School of Mines
Golden
United States of America

Editorial Office
MDPI
St. Alban-Anlage 66
4052 Basel, Switzerland

This is a reprint of articles from the Special Issue published online in the open access journal *Mining* (ISSN 2673-6489) (available at: https://www.mdpi.com/si/mining/future_mining).

For citation purposes, cite each article independently as indicated on the article page online and as indicated below:

Lastname, A.A.; Lastname, B.B. Article Title. *Journal Name* **Year**, *Volume Number*, Page Range.

ISBN 978-3-0365-9941-0 (Hbk)
ISBN 978-3-0365-9942-7 (PDF)
doi.org/10.3390/books978-3-0365-9942-7

Contents

mining

MDPI

Editorial

Editorial for Special Issue "Envisioning the Future of Mining"

Juan M. Menéndez-Aguado [1,*], Oscar Jaime Restrepo Baena [2] and Jessica M. Smith [3]

1 Asturias Raw Materials Institute, Universidad de Oviedo, 33600 Mieres, Spain
2 Department of Materials and Minerals, School of Mines, Universidad Nacional de Colombia, Medellin 65-223, Colombia; ojrestre@unal.edu.co
3 Department of Engineering, Design & Society, Colorado School of Mines, Golden, CO 80401, USA; jmsmith@mines.edu
* Correspondence: maguado@uniovi.es

1. Introduction and Scope

According to the International Energy Agency, clean energy transitions significantly increase strategic minerals demand. As evidence suggests, from 2017 to 2022, overall demand for lithium increased by 300%, 70% for cobalt, and 40% for nickel. Regarding the forecasts, under the Net-Zero emissions by 2050 Scenario, critical mineral production is projected to increase by 350% of the values reached in 2030 [1] unless there are significant changes in urban infrastructure, transportation, and everyday practices [2]. Industry observers caution that a pervasive sense of urgency to meet the growing mineral demands of the energy transition can deepen social and environmental injustices if proper engagement procedures such as Free, Prior and Informed Consent are not followed, as a majority of potential new mineral developments are located on or near land held by Indigenous and land-dependent people [3].

Producing mineral raw materials has faced many challenges in providing the necessary supplies to almost any production chain. In addition to the traditional search for more efficient processes, cleaner and safer operations, and higher levels of community benefit and social acceptance, there is a need for the mining industry to become more sustainable, aiming to become a significant factor in the circular economy, decarbonization, and digital transformation processes. The articles in this Special Issue "Envisioning the Future of Mining" advance our knowledge of the interlinked technical, environmental, and social challenges facing the sector.

Citation: Menéndez-Aguado, J.M.; Restrepo Baena, O.J.; Smith, J.M. Editorial for Special Issue "Envisioning the Future of Mining". *Mining* **2024**, *4*, 1–6. https://doi.org/10.3390/mining4010001

Received: 14 December 2023
Accepted: 19 December 2023
Published: 21 December 2023

2. The Challenges in Future Mining

The trends of recent decades Indicate that current and future scientific and technological development will be marked by what could be called the "Era of Technological Convergence" [4]. Since the middle of the 20th century, a phenomenon of integration between different sciences and technologies has been taking place on an ever-increasing scale. Fields of science and technology that in past times did not seem to have any apparent relationship are now the protagonists of an unprecedented interaction that is shaping a new scientific–technological paradigm. This conception of research work makes it possible to address and attempt to solve complex problems, which are systemic in nature and common to different areas of knowledge, through inter-, multi- and transdisciplinary cooperation.

The term converging technologies was first used by researchers Roco and Bainbridge, who were the editors of the report "Converging Technologies for Improving Human Performance" [4]. For Roco and Bainbridge, the term converging technologies refers to the synergistic combination of four strategic areas of science and technology, each of which continues to progress on its own at an accelerated pace: (1) Nanoscience and nanotechnology; (2) Biotechnology and biomedicine, including genetic engineering; (3) Information technologies, including advanced computing and communication; and (4) Cognitive sciences, including neurosciences. To express this integration of approaches and knowledge in simplified form, the acronym NBIC (Nano, Bio, Info, Cogno) is used. The distinctive character of converging technologies lies in the strong complementarity between them in the study and the possibilities of controlled manipulation of interactions between living and artificial systems. The basic units of study that are fundamental to all converging technologies originate at the nanoscale.

Yet there is a high risk that the societal transformations driven by the convergence of physical, digital and biological technologies will lead to exaggerated technological optimism and transhumanist visions. In this case, it is important to insist that new technologies ought to adjust to the needs of society rather than society adapt to the progress of technologies. In this sense, the discussion on technological development should move beyond its technical feasibility and debate on the potential ethical, moral and social implications and limitations in the medium and long term.

Mining activity is not indifferent to these transformations and must be carefully considered in the future.

Technical challenges are top of mind in ensuring an adequate supply to the predicted exponentially growing demand. The *Technology and Innovation challenge* refers to the sustainable mining requirement of adopting new technologies and practices, which can be costly and require significant research and development. Often, these new technologies and practices are responses to environmental goals. For example, the *Energy Consumption challenge* addresses the vast amount of energy mining operations need, often from non-renewable sources. A transition to sustainable energy sources is occurring on many mine sites, and more could be spurred by recent multi-million-dollar investments in clean energy demonstrations on minelands by the United States Department of Energy.

Mining activity brings environmental change that must be better managed. The consumption of vast quantities of water in mining, which can lead to local water scarcity and contamination, constitutes the *Water Management challenge*. Sustainable mining practices need to reduce water consumption and manage it responsibly.

Not properly conducting mining operations can lead to deforestation, soil erosion, water pollution, and habitat destruction. The *Environmental Impact challenge* refers to sustainable mitigation through responsible land reclamation, reduced waste generation, and adequate water management. The *Waste Management challenge* is becoming ever more pressing, as mines generate larger and larger amounts of waste, including tailings and slag, as the ore grades of remaining mineral resources are generally lower than those of previous and current mines. Some companies are addressing this challenge by finding ways to reprocess tailings to recover more minerals. Finally, the *Biodiversity Conservation challenge* tackles the risk of disruption of local ecosystems and biodiversity threats due to improper management of mining operations. Sustainable mining should consider the protection and restoration of affected ecosystems. Without proper management, environmental problems can quickly become social conflicts [5].

Community challenges address the potential impact of mining operations on local communities, including displacement, loss of livelihoods, and health concerns [6]. Sustainable mining involves community consent and engagement, fair labour practices, and benefit-sharing agreements. Mining operations often face resistance from environmental groups and the general public, as is evident in the growing difficulties of permitting new mines, so gaining and maintaining social acceptance and trust is an ongoing challenge for sustainable mining. While many frame this challenge as one of public perception, which focuses on changing opinions about mining, this challenge could be reframed as one of alignment, which would instead focus on designing and operating mines in ways that are consistent with local expectations and values [7,8]. The *Legal and Regulatory Compliance challenge* signals that sustainable mining requires navigating complex and continually changing legal and regulatory frameworks, often across national borders. Lastly, the *Human Rights challenge* is essential: protecting the rights of workers and local communities is a fundamental aspect of sustainable mining. Ensuring fair labour practices avoiding human rights violations presently and in the future has no possible debate. The sense of urgency to plan and permit new mines cannot trump the protection of human rights.

The *Carbon neutral operations challenge* is ramping up in importance. With this, efforts are intensifying to find new materials that can totally or partially replace those that are traditionally exploited, using CO_2 capture methods, replacing traditional fossil fuels with alternative fuels and increasing efficiency in the production process, as well as seeking new uses and opportunities for traditional materials and their wastes by developing materials that are increasingly durable over time, among other impact measures.

In line with this objective and following the needs expressed in the industry, the importance of promoting the development of projects focused on the circular economy within the mining industry, within the framework of environmental sustainability, is recognised. The objective is based on promoting the adoption of circular practices in mining by identifying opportunities to reduce waste or generate new products from it, reuse materials and optimise extraction and production processes. Through research and collaboration with different actors in the sector, we seek to create innovative solutions that contribute to minimising the environmental impact of mining and preserving natural resources. The potential of the circular economy is considered fundamental to positively transform the mining industry and thus encourage its implementation through projects that foster sustainability and promote a responsible and efficient approach to the use of resources.

Finally, we can consider the *Global Supply Chain challenge* [9]. As mining operations are often part of complex global supply chains, ensuring responsible sourcing of minerals, traceability and transparency throughout the supply chain can be challenging but necessary [10]. The practice of artisanal and small-scale mining, and urban mining, are central to this challenge, as these activities are often performed informally, outside of the view of states and the private industry.

In the European context, the recently presented Raw Materials Act [11] proposes a regulatory framework designed to address the challenges faced by the European Union in the strategic sectors of decarbonisation, digitalisation, and aerospace and defence. The proposal establishes benchmarks for minimum shares of E.U. demand to be covered by domestically produced and recycled raw materials. Also, it aims to reduce dependencies on single third-country suppliers in all supply chain steps, stressing the importance of increasing supply security and sustainability through circularity, standardisation efforts, skill development, and strategic actions for research and innovation [12].

3. An Overview of the Published Articles

The Aguayo et al. article (Contribution 1) addresses the Technology and Innovation Challenge, discussing the potential productivity and safety benefits that incorporating a surge loader may bring to the load and haul system by analysing the system, component characteristics, and mine planning aspects. With the available data on the operation of this equipment and the incident data from Chile and Peru, they point out that the surge loader addition to the shovel–truck system is an innovation that can improve both the productivity and the safety of the loading and hauling activities.

The article by Afolayan et al. (Contribution 2) focuses more on the Community Challenge, dealing with health and safety issues and legal and regulatory ones in the case of barite mining in Nigeria. The exposure of artisanal miners to polluted air, water, and soil is thoroughly evaluated. Some recommendations are presented on the need for annual medical outreach to mining sites and the use of technology (AI) for future mining.

Contribution 3 (Young and Rogers) revisits the Technology and Innovation Challenge on mine hauling, focusing on dumping operations and proposing a method for generating high-fidelity models of dump profiles. They develop photogrammetric models of dumps using unmanned aerial vehicles with mounted cameras. The research identify the factors that influence these profiles, mainly the truck's location relative to the dump crest, the movement of the underlying dump material during the dumping process, and the differences in the dump profile before dumping.

Continuing with the same technological challenge, Amoako et al. (Contribution 4) introduce machine learning algorithms to model rock fragmentation in mine blasting operations. The paper successfully demonstrates the potential of achieving higher accuracy in mean rock fragment size prediction using a multilayered artificial neural network and support vector regression, improving the conventional Kuznetsov empirical model. The trained models could be incorporated into existing fragmentation analysis software to provide blasting engineers with more accurate estimations.

Contribution 5 (Mammadli et al.) also addresses computational tools, but in this case, the analysis focuses on evaluating co- and by-products. The proposed methodology is applied to assess the production status of different commodities in a polymetallic deposit located in Azerbaijan. The evaluation outcomes quantify the production potentials for several commodities in the deposit. The authors justify using this tool to evaluate all kinds of polymetallic deposits concerning the co- and by-production of several minor critical raw materials.

In the case of Contribution 6 (Talebi et al.), the focus leaps again to health and safety issues, but this time using advanced I.T. procedures. In particular, the paper provides an approach to using operational data sets to find the leading indicators of truck operators' fatigue. A machine learning algorithm is used to model the individual's fatigue, and a model is proposed with the algorithm and an extensive data set. The results show that the model can find the importance of the individual factors along with work and environmental factors among operational data sets.

Bao et al. (Contribution 7) review the electrification alternatives for open pit mine haulage, facing one of the most significant challenges posed by the net zero emissions target to the mining sector. In the paper, the authors examine options for decarbonising the haulage systems in large surface mines, comparing electrification alternatives for large surface mines, including In-Pit Crushing and Conveying (IPCC), Trolley Assist (T.A.) and Battery Trolley (B.T.) systems. These emerging technologies provide mining companies and associated industries with opportunities to adopt zero-emission solutions and help transition to an intelligent electric mining future.

The Schlezak and Styer article (Contribution 8) directly addresses the Community Challenge with the proposal of the *inclusive urban mining* concept. They illustrate that inclusiveness and the circular economy can come together in new forms of urban mining, analysing the cases of construction and demolition waste and e-waste sectors in Colombia and Argentina from a sociotechnical perspective. As a result, they highlight the importance of promoting community-based research methods and concepts to be included in mining, materials, metallurgical science, and engineering academic programs to address these challenges.

Contribution 9 (Smith et al.) stresses the importance of a sociotechnical approach to future engineers of natural resources to understand and promote social justice and sustainability in professional development. The future changes that current challenges will produce need the engineer contribution and promotion as active parts of society. This research is carried out with two different groups of engineering students from the Colorado School of Mines and the Universidad Nacional de Colombia. The researchers find that collaborative, interdisciplinary teaching about authentic problems enhances student abilities to understand their professions from a sociotechnical perspective.

Finally, Contribution 10 (El Hiouile et al.) presents a case study of the application of artificial intelligence to monitor a screen unit in a phosphate processing plant. Using artificial intelligence and image processing techniques, this research evaluates the performance of machine learning and deep learning models to detect the screening unit malfunction in the open pit of the Benguerir phosphate mine in Morocco. The results prove the robustness of models based on convolutional neural networks (CNN) and the Histogram of Oriented Gradient (HOG) technique.

4. Conclusions

Under the sustainable development principles framework, further insight must be gained to overcome the potential challenges in mineral raw material production [13]. This issue, "Envisioning the Future of Mining", covers new sources of raw materials (urban mining, deep sea mining, ultradeep mining, extraterrestrial mining), the continuously growing levels of digitalisation and automation, and the use of safer, healthier and cleaner technologies in raw material processing and extracting. It also spans or scales from artisanal to large-scale mining activities and includes perspectives from education research that point the way to training the next generation of industry professionals to address these and other challenges in sustainable and socially responsible ways.

Conflicts of Interest: The authors declare no conflict of interest.

List of Contributions

1. Aguayo, I.; Nehring, M.; Ullah, G. Optimising Productivity and Safety of the Open Pit Loading and Haulage System with a Surge Loader. *Mining* **2021**, *1*, 167–179. https://doi.org/10.3390/mining1020011.
2. Afolayan, D.; Onwualu, A.; Eggleston, C.; Adetunji, A.; Tao, M.; Amankwah, R. Safe Mining Assessment of Artisanal Barite Mining Activities in Nigeria. *Mining* **2021**, *1*, 224–240. https://doi.org/10.3390/mining1020015.
3. Young, A.; Rogers, W. A High-Fidelity Modelling Method for Mine Haul Truck Dumping Process. *Mining* **2022**, *2*, 86–102. https://doi.org/10.3390/mining2010006.
4. Amoako, R.; Jha, A.; Zhong, S. Rock Fragmentation Prediction Using an Artificial Neural Network and Support Vector Regression Hybrid Approach. *Mining* **2022**, *2*, 233–247. https://doi.org/10.3390/mining2020013.
5. Mammadli, A.; Barakos, G.; Islam, M.; Mischo, H.; Hitch, M. Development of a Smart Computational Tool for the Evaluation of Co- and By-Products in Mining Projects Using Chovdar Gold Ore Deposit in Azerbaijan as a Case Study. *Mining* **2022**, *2*, 487–510. https://doi.org/10.3390/mining2030026.
6. Talebi, E.; Rogers, W.; Drews, F. Environmental and Work Factors That Drive Fatigue of Individual Haul Truck Drivers. *Mining* **2022**, *2*, 542–565. https://doi.org/10.3390/mining2030029.
7. Bao, H.; Knights, P.; Kizil, M.; Nehring, M. Electrification Alternatives for Open Pit Mine Haulage. *Mining* **2023**, *3*, 1–25. https://doi.org/10.3390/mining3010001.
8. Schlezak, S.; Styer, J. Inclusive Urban Mining: An Opportunity for Engineering Education. *Mining* **2023**, *3*, 284–303. https://doi.org/10.3390/mining3020018.
9. Smith, J.; McClelland, C.; Restrepo, O. Sociotechnical Undergraduate Education for the Future of Natural Resource Production. *Mining* **2023**, *3*, 387–398. https://doi.org/10.3390/mining3020023.
10. El Hiouile, L.; Errami, A.; Azami, N. Toward Automatic Monitoring for Anomaly Detection in Open-Pit Phosphate Mines Using Artificial Vision: A Case Study of the Screening Unit. *Mining* **2023**, *3*, 645–658. https://doi.org/10.3390/mining3040035.

References

1. IEA. Available online: https://www.iea.org/topics/critical-minerals (accessed on 1 December 2023).
2. Riofrancos, T.; Kendall, A.; Dayemo, K.; Haugen, M.; McDonald, K.; Hassan, B.; Slattery, M.; Xan, L.; Achieving Zero Emissions with More Mobility and Less Mining. Climate and Community Project. 2023. Available online: http://www.climateandcommunity.org/more-mobility-less-mining (accessed on 31 January 2023).
3. Owen, J.R.; Kemp, D.; Harris, J.; Lechner, A.M.; Lèbre, É. Fast track to failure? Energy transition minerals and the future of consultation and consent. *Energy Res. Soc. Sci.* **2022**, *89*, 102665. [CrossRef]
4. Roco, M.C.; Bainbridge, W.S. *Converging Technologies for Improving Human Performance Nanotechnology, Biotechnology, Information Technology and Cognitive Science NSF/DOC-Sponsored Report*; National Science Foundation: Arlington, VA, USA, 2002.
5. Franks, D.M.; Davis, R.; Bebbington, A.J.; Ali, S.H.; Kemp, D.; Scurrah, M. Conflict translates environmental and social risk into business costs. *Proc. Natl. Acad. Sci. USA* **2014**, *111*, 7576–7581. [CrossRef] [PubMed]

6. Payne Institute for Public Policy. The State of Critical Minerals Report. *Colorado School of Mines.* 2023. Available online: https://payneinstitute.mines.edu/wp-content/uploads/sites/149/2023/09/Payne-Institute-The-State-of-Critical-Minerals-Report-2023.pdf (accessed on 1 December 2023).
7. Owen, J.R.; Kemp, D. Social license and mining: A critical perspective. *Resour. Policy* **2013**, *38*, 29–35. [CrossRef]
8. Smith, J.M. *Extracting Accountability: Engineers and Corporate Social Responsibility*; The MIT Press: Cambridge, MA, USA, 2021.
9. Kondratev, V.B.; Popov, V.V.; Kedrova, G.V. Critical Materials' Supply Chains and US National Security. *World Econ. Int. Relat.* **2023**, *67*, 5–16. [CrossRef]
10. Franks, D.M.; Keenan, J.; Hailu, D. Mineral security essential to achieving the Sustainable Development Goals Nature. *Sustainability* **2023**, *6*, 21–27.
11. European Critical Raw Materials Act(a). European Commission–European Commission. 2023. Available online: https://ec.europa.eu/commission/presscorner/detail/en/ip_23_1661 (accessed on 11 July 2023).
12. Hool, A.; Helbig, C.; Wierink, G. Challenges and opportunities of the European Critical Raw Materials Act. *Miner. Econ.* **2023**. [CrossRef]
13. Ericsson, M. The evolving structure of the global mining industry. *Mater. Et Tech.* **2023**, *111*, 303. [CrossRef]

mining

MDPI

Article

Optimising Productivity and Safety of the Open Pit Loading and Haulage System with a Surge Loader

Ignacio Andrés Osses Aguayo [1], Micah Nehring [2,*] and G. M. Wali Ullah [3]

[1] Ingeniera Civil de Minas, Universidad de Concepción, Concepción 3349001, Chile; igosses@udec.cl
[2] School of Mechanical and Mining Engineering, The University of Queensland, Brisbane, QLD 4072, Australia
[3] Department of Mathematics, University of Chittagong, Chattogram 4331, Bangladesh; wali@cu.ac.bd
* Correspondence: m.nehring@uq.edu.au

Abstract: The open pit mining load and haul system has been a mainstay of the mining industry for many years. While machines have increased in size and scale and automation has become an important development, there have been few innovations to the actual load and haul process itself in recent times. This research highlights some of the potential productivity and safety benefits that the incorporation of a surge loader may bring to the load and haul system through an analysis of the system, discussion of component characteristics, and mine planning aspects. The incorporation of the surge loader into open pit loading and haulage operations also enables improved safety. This is a result of a reduction in shovel–truck interactions and the reduced likelihood of truck overfilling and uneven loading. This paper details the number of mine worker deaths that a surge loader may have prevented within the Peruvian and Chilean mining industries.

Keywords: surge loader; mine safety; load and haul; truck and shovel; open pit mining; haulage systems

Citation: Aguayo, I.A.O.; Nehring, M.; Ullah, G..M..W. Optimising Productivity and Safety of the Open Pit Loading and Haulage System with a Surge Loader. *Mining* **2021**, *1*, 167–179. https://doi.org/10.3390/mining1020011

Academic Editors: Juan M. Menéndez-Aguado, Oscar Jaime Restrepo Baena and Jessica M. Smith

Received: 13 June 2021
Accepted: 26 July 2021
Published: 2 August 2021

1. Introduction

The minerals industry is required to process increasingly lower grade ores [1] in order to meet an insatiable demand for raw materials that is being driven by the general advancement of humanity, and in particular, the global transition toward cleaner forms of energy and its respective storage and transmission. This is occurring against a backdrop of heightened community awareness of environmental, social, and governance issues that have sadly plagued the industry for too long. To meet the challenges associated with having to extract and process greater quantities of ore material at a price that remains reasonable, it is necessary to increase the productivity of operations [2] through innovations.

The safety of operations is essential in the mining industry [3]. This is why the objective of any innovation should also be to increase the safety of operations while lifting their productivity at the same time [3]. Even though it may be challenging, this is the key reason why innovation in the mining industry needs to look beyond small-scale incremental improvements of current systems. Step-change innovation will only occur if new ways to redesign the various stages of the operation are achieved [3].

In current open pit mining operations, one of the highest costs lies in the loading and hauling stage [4]. During the loading and hauling stage, material at the working face is loaded by an excavator/shovel into trucks, which have positioned themselves as best as possible to receive this material. The truck then proceeds to haul this material to either a processing plant, waste dump, or stockpile. Although the current Shovel-Truck (ST) system presents numerous advantages, particularly over inflexible conveyor-based continuous mining systems, it is nevertheless costly and becomes more cost as the operation matures. For this reason, it is vital for the future of open pit operations to find ways to improve the Shovel-Truck system. In this context, the incorporation of a surge loader into this system is worthy of further investigation.

This research highlights some of the potential productivity and safety benefits that the incorporation of a surge loader may bring to the load and haul system through an analysis of the system, discussion of component characteristics, mine planning, and design aspects. While the authors are aware of some in-house studies conducted by various mining companies into the use of surge loaders, very few studies are available in the open literature. This paper thus presents a detailed discussion into the potential productivity and safety benefits that the introduction of a surge loader into the open pit loading and haulage system may present. This is the first of a series of papers that the authors envisage will ultimately delve in-depth into the technical aspects of the use of a surge loader.

2. Background

The open pit mining value chain is comprised of initial prospecting and exploration, resource modelling and mine planning, mine production including drilling and blasting followed by loading and hauling, comminution to liberate and separate the valuable mineral, followed by further refining, and finally, transportation to market [5]. In open pit mining operations, the loading and hauling stage is very important in the overall production of the mine. This is because the performance in this stage largely determines the production rate that the mining operation can achieve. Often the loading and hauling stage is the limitation or bottleneck across the whole open pit mining value chain [6]. As such, an efficient, safe, and well-functioning loading and hauling system is essential to maximising mine productivity [7,8] and value for all stakeholders.

The purpose of the loading and hauling stage is to move the material previously fragmented by the drilling and blasting process. The first step in this process consists of loading the material from the bench or working face of the mine into trucks. The next step involves transporting this material to its destination (stockpile, waste dump, or processing plant), via a haul road that generally spirals up the pit walls and is specially designed and maintained to accommodate large haulage trucks of up to a 400-tonne capacity [9].

Depending on operational characteristics and site geometry, the loading and hauling of material typically represent between 35–55% of the operational costs of an open pit mine [10,11]. As open pit mining operations mature and additional resources are discovered, extensions to mine life are common. This results in greater pit depths and thus longer hauls. Ultimately, the cost of material transportation takes a larger and larger share of the operational costs of the site [6,11]. For this reason, one of the challenges for open pit mining operations is to continually optimize the loading and hauling stage. Any improvement could ultimately mean being able to extract additional ore at a greater depth, which would lead to a further increase in the useful life of the open pit mining operation [5].

3. Shovel-Truck System

Currently, the most common system for the loading and hauling stage in open pit mining is the Shovel-Truck system (ST) [9,12]. This comprises of a shovel loading blasted material into a truck, which transports the material from the dig face to a destination (stockpile, dump, or primary crusher) where it is unloaded. The truck then returns to the shovel and the cycle repeats, as shown in Figure 1. This has been the mainstay of open pit mining for many years. This system is simple and easy to implement in mining operations and is considered as being very reliable, flexible, and effective [9,13]. One of the main advantages of this system is its versatility as it only needs roads suitable for the movement of trucks without the need for more complex infrastructure such as conveyors. A substitute truck may also be rapidly dispatched to replace any breakdown. This allows the system to easily change and fit to the design of the mine and is thus more likely to prevent the Shovel-Truck system from becoming a limitation in any mine production expansion [14,15].

The Shovel-Truck system has undergone multiple improvements over time. One of these improvements involves optimization of the travel route, whereby the most efficient route for the system is selected through production and topographic analysis. This reduces cycle time, transport cost, and increases the productivity of the system [16,17]. All these changes have improved the productivity of the system as a whole, however, they have not altered the cycle itself (Figure 1) and thus only generate incremental improvements.

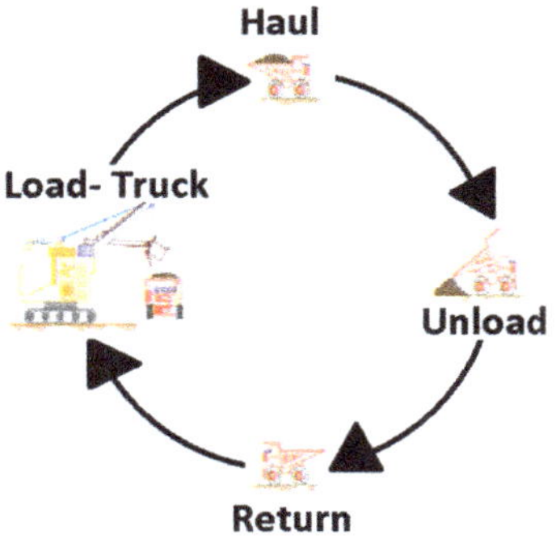

Figure 1. Cycle of the Shovel-Truck system.

Although the Shovel-Truck system has proven its functionality over the years, some safety aspects remain [18–20]. The various interactions between the shovel and the truck also lack precise control over the truck's fill factor.

4. Surge Loader System

While the concept of the surge loader has been in existence for many years, it is only recently that an equipment manufacturer is offering this as a standard product to the broader traditional open pit mining industry. In conjunction with the further development of scanning and sensing technologies, the latest surge loader now has additional capabilities that it previously did not. This now makes the surge loader a new and exciting proposition for many open pit mining operations around the world. As with any new potential equipment purchase, a thorough study should be conducted to understand if the ongoing additional revenue as a result of productivity improvements is able to offset the initial substantial capital cost associated with the purchase and commissioning of a surge loader.

In order to introduce this piece of equipment and due to a lack of alternatives, this paper contains illustrations mostly of the 'Fully Mobile Surge Loader' from MMD. The MMD Fully Mobile Surge Loader is the first of its kind in the world. It is designed to revolutionize the loading of haul trucks; making the process faster, more efficient, and safer. It should be noted that the authors have no affiliation with this product or its manufacturer.

The surge loader is a device designed for use in the loading process, whose main function is to receive the material from the shovel and to then load trucks [21]. It thus serves as an intermediary between the shovel and the truck. The result is an increase in the safety and productivity of the loading and hauling stage, as it divides the cycle of the Shovel-Truck system into two independent cycles [21] as shown in Figure 2.

The cycle of the Shovel-Truck system is divided into two cycles as a result of the surge loader. This means that the shovel and the truck no longer directly interact. Rather, the surge loader acts as an intermediary, which allows functional independence between the loading machine and trucks, respectively.

The surge loader consists of a hopper with a capacity that is generally about 2.5 times (but not always) the capacity of the truck used in the mining operation [22]. The hopper receives material from the shovel which it then transfers to a truck, as shown in Figure 3. With the surge loader, the shovel no longer depends on the immediate presence of a truck to complete its cycle. The continued operation of the shovel now only depends on the remaining available capacity of the hopper. This eliminates direct dependence between the shovel and the truck that exists in the classic Shovel-Truck system [23]. The hopper is track mounted which gives it the mobility and ability to be completely autonomous [23] and to accompany the shovel under normal bench operating conditions [21]. This is a great advantage when compared to other loading and transport systems used in open pit mining with limited mobility and autonomy, such as In-pit Crusher and Conveyor (IPCC) systems [24]. However, it must also be recognized that the surge loader is an additional item of equipment that may breakdown and require maintenance. While planned maintenance may take place to minimize the disruption to productivity as much as possible, any unplanned maintenance as a result of a breakdown is likely to cause a larger disruption to production than it otherwise would.

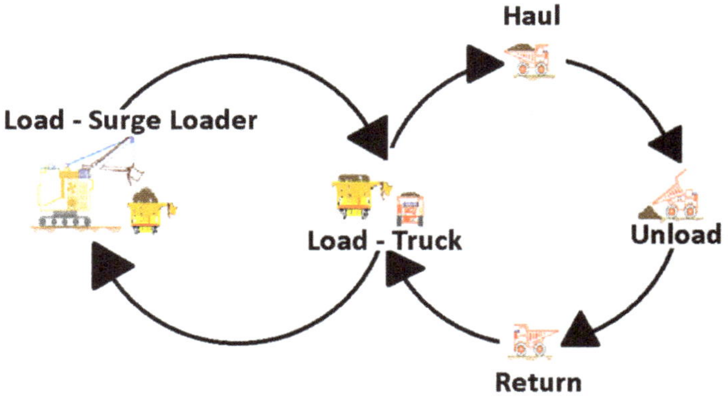

Figure 2. New Shovel-Truck system.

Figure 3. Fully Mobile Surge Loader [21].

4.1. Feeding System

The surge loader feeding system utilises a 'Heavy Duty Apron Plate Feeder' [23]. Figure 4 shows a surge loader feeding system in operation within a coal mine in Colombia. This installation has made it possible for the manufacturer to test its functionality under different operational scenarios, in addition to proving its efficiency in the loading process [23].

The feeding system consists of a conveyor belt that carries the material from the hopper to the truck. The use of the conveyor belt allows the average filling of 330-tonne trucks in 60 s [23]. This represents a significant reduction of approximately 50% in the time taken to load trucks compared to the traditional Shovel-Truck system [1].

Figure 4. Feeding and Loading equipment (Colombia) [23].

The feeding system is comprised of high resistance plates, designed to handle high impact and abrasive materials [25]. These plates are designed for low maintenance requirements, long useful life, and robustness and reliability. They feature overlapping edges which prevent spillage between the plates and are fixed to chains with bolts that are positioned between the grousers, protecting the bolt heads from damage caused by the material being conveyed as illustrated in Figure 5 [25].

Figure 5. Conveyor belt [25].

The main features of the apron plate feeder are the heavy-duty chains and rollers (manufactured by Caterpillar as part of the MMD system), which are attached to the main frame, as shown in Figure 6. These stand out for their high resistance and elimination of impact energy, which is initially absorbed by the conveyor plates, by deforming within their elastic limits. The impact rails then transmit the forces which are dissipated into the main frame construction.

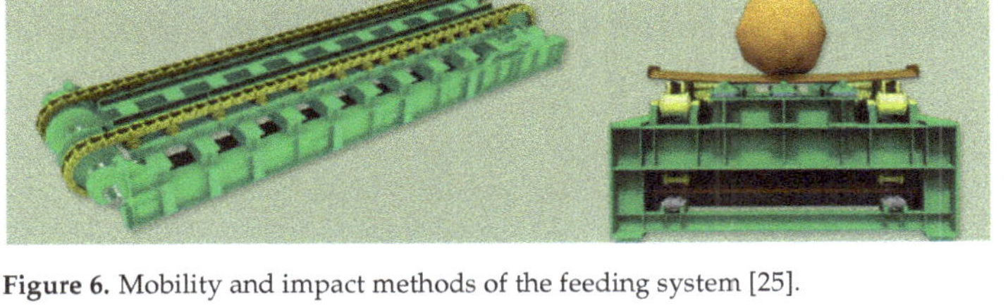

Figure 6. Mobility and impact methods of the feeding system [25].

4.2. Receiving and Loading of the Material

The transfer of material between the surge loader and the truck is carried out through a conveyor belt that carries the material previously deposited into the hopper by the shovel to the truck as shown in Figure 7. The belt, through strategically placed sensors, is capable of measuring the volume and weight of the material that is being delivered, managing to accurately control the fill factor of the truck [20,21]. Through these sensors, the surge loader is capable of detecting the presence of large rocks which can affect how this material is transported [23,26,27]. Together with the information of the material already loaded into the truck, this allows the monitoring system to decide if the truck is capable of transporting this larger rock without exceeding safety limitations [23]. In this case, if the result is negative, the surge loader stops loading and gives the signal for the truck to continue its cycle with the material that was already loaded. In addition to this, the system leaves enough material between the large rock and the material discharge point to serve as a cushion for the impact of the large rock discharging (Figure 8) into the truck tray. The on-board sensor system gives the surge loader a great advantage over the conventional material transfer system (Shovel-Truck system), by accurately controlling the fill factor of the truck and significantly narrowing the fill factor variance [22]. Controlling the fill factor in the classic system (Shovel-Truck system) is carried out by appropriately matching the capacities of the shovel and the truck. Fill factors in the classic system are also influenced by manoeuvrability, the distribution of the material in the shovel bucket, material fragmentation, and operator competence [28]. The effects of many of these items are diminished with the incorporation of a surge loader.

Figure 7. Material transfer sequence [23].

Figure 8. Presence of a large rock on the conveyor belt [23].

From a review of its components, it is evident that a surge loader is a large piece of equipment and requires a significant geometrical footprint. As a result of this, the mine design and subsequent plan may have to be altered in order to accommodate this machine.

5. Effect of the Surge Loader on Truck Productivity

In the classic Shovel-Truck system (without the inclusion of the surge loader) the truck fill factor depends entirely on the shovel and how much material it can load into the truck. This generates three possible scenarios. The first is when the truck exceeds its payload (overfill), which generates an increase in the risk associated with the transport of material [18–20]. For this reason, the on-board measurement system seeks to avoid this and promptly cuts the material supply from the transfer conveyor. The second scenario is when the opposite occurs and the truck is loaded with less than its payload (underfill). This scenario is the most common and causes a decrease in the designed productivity of the truck. The third scenario is where the truck is loaded with its exact payload capacity. This scenario is optimal, but due to different factors such as the loading capacity of the shovel or the competence of the operator, it is very difficult to obtain. This makes the choice of shovels and trucks dependent on each other from a productivity viewpoint. Shovels that can load trucks in as few cycles as possible is thus favoured as it minimises the deviation in the truck fill factor.

The surge loader, through its feeding system, allows for the controlled loading of material into the truck. This is not only practical from the point of view of reducing the loading time, but the feeding system also allows for the control of the fill factor, which is now not dependant on the loading capacity of the shovel but rather on the payload of the truck. This makes it possible to consistently achieve filling factors as close to 100% as possible [23]. This, in turn, allows operations to close the gap on the ideal scenario, which is to achieve maximum truck productivity safely and without increasing the cost of haulage. Given that the truck fill factor of the classic Shovel-Truck system tends to be approximately 90% on average [29], raising this to close to 100% represents a significant potential improvement in productivity for the same number of trucks. In some cases, this productivity improvement may even require a smaller trucking fleet. For other operations, improved trucking productivity may change the production bottleneck from the mine to another aspect of the operation.

The use of the surge loader also results in shovels and trucks operating independently of each other, since the surge loader eliminates the interaction between this equipment. This independence potentially allows for a greater range of shovels and trucks that could be utilized in the operation. This allows the option of having different types and capacities of trucks operating together to undertake the hauling of material, which allows for a more dynamic loading and hauling stage that may better adapt to the different types of materials in the mine at the various stages of extraction.

6. Safety Features

In recent years, great efforts have been made to improve the health and safety of mine workers through innovation in the methods and machines used in mining operations [3,30,31].

6.1. Sensors and Cameras System

The use of autonomous equipment is increasingly common in mining. The surge loader [3] can also operate autonomously. For autonomous vehicles to operate successfully they need to be aware of the distance between themselves and other vehicles with enough time to make safe and reliable mission plans [32,33]. The use of cameras and sensors are therefore not only essential for the correct operation of these vehicles and equipment but to also maintain the high standards of safety necessary in the mining industry [18,20].

The surge loader uses a network of sensors and cameras to detect their surroundings including the approach and departure of trucks. This system determines when a truck is approaching the loading point as shown in Figure 9. Then, using Radio Frequency Identification (RFID) (RFID uses electromagnetic fields to automatically identify and track tags attached to objects) sensors, it is capable of detecting when the truck is at the exact loading point required for the surge loader. As shown in Figure 10, a signal is sent to the truck to stop. Finally, the surge loader returns a signal to the truck when it is loaded to continue its journey. This detection system is completely computerized and both cameras and sensors are remotely controlled. This allows for the material transport process to be automated and eliminates the need for a human operator inside the truck to determine the stopping point.

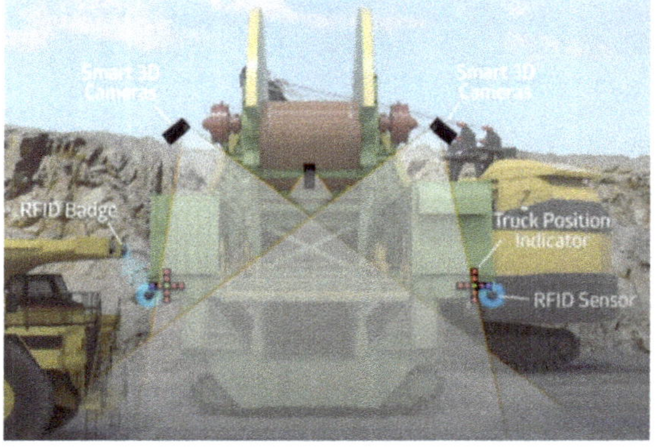

Figure 9. Camera system [23].

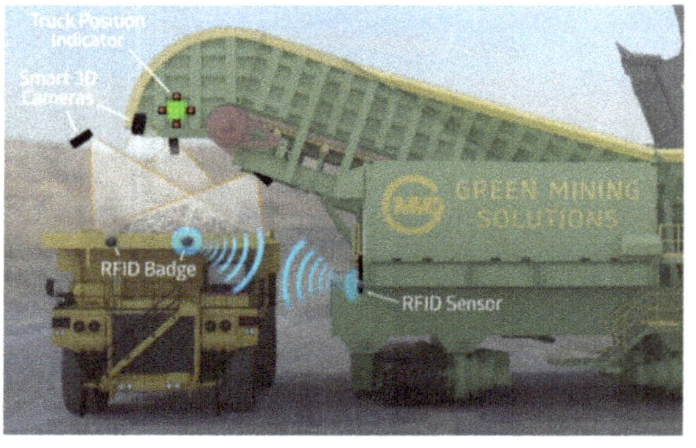

Figure 10. Sensor system [23].

While the use of sensor and scanning technology is no doubt having profound impacts across many industries, the environment in which these are used in mining operations is often prone to dust, which may hinder the full capability of some sensors and scanners.

6.2. Travel Routes

The routes that trucks take in the material transport process of the Shovel-Truck system are an essential aspect since they influence the productivity achieved by the truck. That is why it is important to work with optimal routes that improve the productivity of the truck [17]. In the classic Shovel-Truck system, however, although truck routes are optimized, there are 'dead' sections where the transportation process is delayed [34] and these sections cannot be eliminated. One of these sections corresponds to the manoeuvres carried out by the truck to locate itself appropriately at the loading point [34].

The surge loader eliminates the need for the truck to maneuver multiple times to be located appropriately near the shovel. This is a very common process in the classic Shovel-Truck system. The use of cameras and sensors mounted on the surge loader facilitates the automation of trucks by enabling an intelligent communication system between the surge loader and the truck. In addition, trucks will take simpler routes (without positioning maneuvers) often by facilitating the drive-by-loading method as shown on the right-hand-side of Figure 11. The inclusion of the surge loader to the Shovel-Truck system thus allows for working with more continuous loading routes. On these new routes, manoeuvring times are largely eliminated. In addition to reducing truck loading time, this reduces the overall truck cycle time.

Figure 11. Truck routes with the surge loader [23].

The proximity detection system of the surge loader reduces operational maneuvers of trucks (turns, reversing, stopping, or exiting the route), due to the ability to detect and eliminate blind spots [32,35]. The decrease in the probability of collisions during normal driving, parking, or maintenance maneuvers is another benefit associated with the detection system. Consequently, this results in extensions of the useful life of rims, tires, and suspension systems [32,35]. The main benefit of the detection system is an increase in safety (by preventing accidents). Other benefits likely include reduced maintenance costs [32,35]. While these benefits are not direct improvements to the loading and hauling stage, they are benefits that affect the mine, its financial viability, and the planning process.

6.3. Safety

Increasing safety is a priority in modern mining operations [19,29,31]. One of the current challenges is to find new ways to achieve greater levels of safety without, or, with minimal reduction, in the productivity of operations [5]. In this context, one of the main risks present in open pit mining is the loading and hauling stage [18–20,36,37], since the Shovel-Truck system (typically used in this stage) requires direct interaction of large equipment [36]. In general, any accident that occurs in the loading and haulage stage results in a delay and decrease in mine productivity [36]. Considering that in open pit mining the loading and haulage stage determines the productivity of the operation [5], it is very important in this type of operation to keep accident rates to a minimum [31,36].

On the world stage in the field of mining, Peru and Chile are both leaders in the production of vital metals including copper, gold, manganese, and zinc. Both have positioned themselves as leaders, which has been achieved through their mining policies, high export rates, and their large number of active mines. Records of fatal accidents occurring in the mining industries of Peru and Chile are tabulated in Tables 1 and 2, respectively.

Table 1 shows the fatal mining accidents registered by the Ministry of Energy and Mines (MINEM) of Peru between 2000 and 2016. It can be seen that 41 accidents out of 842 correspond to accidents that occurred in the loading and haulage stage.

While the number of accidents may seem high it should be noted that most of the registered accidents are concentrated in the first seven years of the sample (2000–2007). As shown in Table 1, the largest source of recorded accidents corresponds to accidents related to falling rocks (representing 32%). Although the loading and hauling stage is not the main source of accidents, this activity represents 5% of the total registered, which is still very important considering that these are accidents involving deaths.

In the case of Chile, the records of the Chilean National Geology and Mining Service (SERNAGEOMIN), observed in Table 2, show that between 2010 and 2019, Chile recorded 228 accidents in the mining industry, of which 41 correspond to accidents that occurred in open pit mining.

Table 1. Number of accidents resulting in death in Peru (2000–2016).

Type of Accident	Number of Accidents
Rock fall	271
Fall of workers	82
Vehicle traffic	76
Others	65
Landslide	63
Intoxication and suffocation	70
Loading and haulage	41
Explosions	33
Equipment maneuvering	54
Electric energy	38
Material handling	19
Burial by subsidence of land	23
Tools	7
Total	842

Source: Ministry of Energy and Mines (MINEM), Government of Peru.

Table 2. Number of accidents resulting in death in Chile (2010–2019).

Location	Number of Accidents
Underground mine	117
Open pit mine	41
Port	2
Workshops	4
Others	7
Road	17
Tailings dump	4
Processing plant	30
Surface installation	6
Total	228

Source: Geology and Mining Service (SERNAGEOMIN), Government of Chile.

Although the highest number of registered accidents corresponds to accidents related to underground mining (more than double the accidents than open pit mining operations), the second-highest source of accidents corresponds to accidents that occurred in open pit mining, representing 18% of the total. As in the case of Peru, the recorded accidents are fatal so this is still very significant.

The various operational mechanisms that the surge loader would negate, including (1) loading the truck without it approaching the shovel, (2) not needing to perform additional maneuvers to accommodate itself, (3) controlling the loading of the material to avoid overfilling, and (4) the transportation of poorly balanced loads due to the presence of large rocks, would increase the safety of the loading and haulage stage of open pit mining operations. Table 3 shows the data in Tables 1 and 2 broken down into further categories. It can be observed that, of the fatal accidents mentioned previously, 27 accidents in the case of Peru could have been avoided. This represents 66% of all accidents related to the loading and haulage stage, with fourteen accidents avoidable with the use of the feeding system, six accidents with the use of the new routes, and seven with the surge loader as an intermediary step (Table 3). In the case of Chile, the use of the surge loader would have prevented seven fatal accidents (Table 3). This number can be perceived as low, but, it represents 17% of the open pit mine accidents, with four accidents being avoidable with the use of the feeding system, one with the use of new routes, and two with the surge loader as an intermediary step. The use of the surge loader could have avoided around 3% of the total fatal accidents in each registry (Peru: 3.21%, Chile: 3.07%). This percentage represents accidents related to truck rollovers on the road because of overfill accidents that occurred due to inappropriate positioning of the truck at the loading point.

Table 3. Avoidable accidents.

Open Pit Mine Accidents	Number Accidents PERU	Number Accidents Chile	
Truck accidents due to overfill	14	4	Avoidable with the use of the surge loader
Truck and Shovel crash accidents	6	1	
Accidents in the process of loading material from the shovel to the truck	7	2	
Not avoidable with the use of the surge loader	14	34	
Total	41	41	

Source: Ministry of Energy and Mines (MINEM), Government of Peru, Geology and Mining Service (SERNAGEOMIN), Government of Chile.

The main aspect of the surge loader that would avoid most of these accidents is its feeding system. This prevents the overfilling of trucks, which, as can be seen from the data recorded in Peru and Chile, is one of the leading reasons for fatal accidents in the loading and haulage stage.

7. Conclusions

In conclusion, adding equipment such as the surge loader to the Shovel-Truck system allows for significant improvement in the production cycle of the loading and hauling stage in open pit mines. The surge loader not only allows for productive independence of both the shovel and the truck but also allows for greater freedom of choice in the selection of trucks. With the incorporation of the surge loader, the shovel and the truck no longer have to maintain a certain match in their capacities. This broadens the catalog of equipment available for the loading and hauling stage.

The surge loader makes it possible to simplify the travel routes of the trucks, which speeds up the transport of material by reducing maneuvering times. In addition, the material unloading system of the surge loader allows for control over the fill factor of the trucks, which increases the efficiency of the loading process. For these reasons, the new cycle that the surge loader applies not only separates the shovel cycle from the truck cycle but also improves the loading and hauling stage.

From a safety point of view, research shows that the different features of the surge loader decrease the risks of the loading and hauling stage by avoiding interactions between the equipment. Its feeding system also manages better control over the fill factor of the trucks than present Shovel-Truck systems. This avoids any possibility of generating an overfill of the truck which increases the risk of accidents as seen in the fatal accident data of Chile and Peru. Its incorporation could have prevented 3% of the fatal mining accidents that occurred in Chile and Peru.

Although there is still very limited data on the operation of this equipment in the field, both its characteristics and the analysis of the accident data from Chile and Peru show that adding the surge loader to the Shovel-Truck system is an innovation that would improve both the productivity and the safety of the loading and hauling stage. The modern surge loader, therefore, represents a potentially new way for future loading and haulage that would be fundamental for the future of open pit mining.

8. Future Research

Several of the aspects of the surge loader requires further investigation. First and foremost, a detailed simulation for the new loading and haulage system that incorporates the surge loader should be undertaken to determine the true impact on productivity. This might consist of a simulation of the shovel productivity and the truck fleet productivity in isolation from each other followed by a simulation that combines the entire loading and haulage fleet.

While conventional thinking suggests that the capacity of the surge loader should be 2.5 times the capacity of the truck, it is yet to be proven in the literature that this is optimal. Another simulation should therefore alter the size of the surge capacity to identify its impact. In addition to this, a mixed truck fleet of varying capacities should also be investigated to determine this impact.

The size and footprint of a surge loader also need to be taken into account in the mine planning and design process. Likely, the deployment of a surge loader onto an operating bench within an open cut mine could warrant the use of a wider pushback. If this is the case, a redesign and of the mine will be required, which will have follow-on impacts on productivity and financial metrics.

While scanning and sensor technology has improved immensely, the impact of dust on the potential disruption of scanners and sensors to prevent adequate communication between equipment should be fully investigated. The dust that can be generated in an open pit mining environment can be significant. It is therefore important that the most suitable type of scanning and sensing technology is used.

While the surge loader reduces several of the typical risks associated with conventional Shovel-Truck loading and haulage, a detailed and thorough investigation should take place to identify any additional risks that the surge loader may introduce into the system. A detailed risk assessment should also determine the consequences and likelihood of these and identify potential mitigation strategies.

Author Contributions: All authors have contributed equally to the conceptualization, literature review, analysis and drafting of this manuscript. All authors have read and agreed to the published version of the manuscript.

Funding: This research received no external funding.

Conflicts of Interest: The authors declare no conflict of interest.

References

1. Dzakpata, I.; Knights, P.; Nehring, M. Truck and Shovel Versus In-Pit Conveyor Systems: A Comparison Of The Valuable Operating Time. In Proceedings of the 16th Coal Operators' Conference, Mining Engineering, Wollongong, NSW, Australia, 10–12 February 2016; pp. 463–476.
2. Paricheh, M.; Osanloo, M. Concurrent open-pit mine production and in-pit crushing—Conveying system planning. *Eng. Optim.* **2020**, *52*, 1780–1795. [CrossRef]
3. Leiva Peña, C.I. Análisis de Implementación de Camiones Autónomos en Mina a cielo Abierto. Bachelor's Thesis, Universidad de Chile, Santiago, Chile, 2020.
4. Zhang, L.; Xia, X. An Integer Programming Approach for Truck-Shovel Dispatching Problem in Open-Pit Mines. *Energy Procedia* **2015**, *75*, 1779–1784. [CrossRef]
5. Hustrulid, W.; Kuchta, M.; Martin, R. *Open Pit Mine Planning & Design*; Student-Pa.; August Aimé Balkema: Rotterdam, The Netherlands, 2013; Volume 1-Fundamen.
6. Dean, M.; Knights, P.; Kizil, M.S.; Nehring, M. Selection and Planning of Fully Mobile IPCC Systems for Deep Open-Pit Metalliferous Applications. *Third Int. Futur. Min. Conf. Proc.* **2015**, 219–223.
7. Berwick, M.; Farooq, M. *Truck Costing Model for Transportation Managers*. 2003. Available online: https://www.ugpti.org/resources/reports/downloads/mpc03-152.pdf (accessed on 8 September 2020).
8. Ozdemir, B.; Kumral, M. Simulation-based optimization of truck-shovel material handling systems in multi-pit surface mines. *Simul. Model. Pract. Theory* **2019**, *95*, 36–48. [CrossRef]
9. Krause, A.J. *Shovel-truck Cycle Simulation Methods in Surface Mining*; University of Witwatersrand: Johannesburg, South Africa, 2006.
10. Alarie, S.; Gamache, M. Overview of solution strategies used in truck dispatching systems for open pit mines. *Int. J. Surf. Min. Reclam. Environ.* **2002**, *16*, 59–76. [CrossRef]

11. Bakhtavar, E.; Mahmoudi, H. Development of a scenario-based robust model for the optimal truck-shovel allocation in open-pit mining. *Comput. Oper. Res.* **2020**, *115*, 104539. [CrossRef]
12. Abbaspour, H.; Drebenstedt, C.; Dindarloo, S.R. Evaluation of safety and social indexes in the selection of transportation system alternatives (Truck-Shovel and IPCCs) in open pit mines. *Saf. Sci.* **2018**, *108*, 1–12. [CrossRef]
13. Abbaspour, H.; Drebenstedt, C.; Paricheh, M.; Ritter, R. Optimum location and relocation plan of semi- mobile in-pit crushing and conveying systems in open-pit mines by transportation problem. *Int. J. Min. Reclam. Environ.* **2019**, *33*, 297–317. [CrossRef]
14. Tapia, E.; Salazar Araya, A.; Saavedra, N.; Nehring, M.; Mora, J. An analysis of full truck versus full bucket strategies in open pit mining loading and hauling operations. *Int. J. Min. Reclam. Environ.* **2021**, *35*, 1–11. [CrossRef]
15. Sharma, H.S. Comparative economic analysis of transportation systems in surface coal mines. *Int. J. Surf. Min. Reclam. Environ.* **1991**, *5*, 17–23.
16. Baek, J.; Choi, Y. A new method for haul road design in open-pit mines to support efficient truck haulage operations. *Appl. Sci.* **2017**, *7*, 747. [CrossRef]
17. Liu, G.; Chai, S. Optimizing open-pit truck route based on minimization of time-varying transport energy consumption. *Math. Probl. Eng.* **2019**, *2019*, 1–12. [CrossRef]
18. Kasap, Y.; Subaşı, E. Risk assessment of occupational groups working in open pit mining: Analytic Hierarchy Process. *J. Sustain. Min.* **2017**, *16*, 38–46. [CrossRef]
19. Teixeira, J.V.S.; Reis, A.M.; Mendes, F.B.; Vergara, L.G.L. *Occupational and Environmental Safety and Health*; Springer: Charm, Switzerland, 2019; Volume 202.
20. Mutual-Seguridad. *Prevención de Riesgos en Minería del Cobre Apunte curso Prevención de Riesgos en Minería del Cobre a rajo Abierto*; Mutual de seguridad CChC: Santiago, Chile, 2018.
21. Support Staff. The Spotlight Shines on Landmark MMD Invention—Prospect Mining Awards. 27 November 2019. Available online: https://prospectawards.com.au/the-spotlight-shines-on-landmark-mmd-invention/ (accessed on 8 September 2020).
22. MMD. *Fully Mobile Surge Loader Catalog*; MMD: Edenvale, South Africa, 2019.
23. McCarthy, P. *MMD Changing the Future of Mining*; Sydney, Australia,, 2019.
24. Londoño, J.G.; Knights, P.F.; Kizil, M.S. Modelling of In-Pit Crusher Conveyor alternatives. *Trans. Institutions Min. Metall. Sect. A Min. Technol.* **2013**, *122*, 193–199. [CrossRef]
25. MMD. *Mmd Sizers*; No. 8.1; El Grupo de Empresas MMD, 2017; Available online: https://www.mmdsizers.com/images/brochure/ES_MMD_Products_and_Applications_Brochure_v8-1-Online.pdf (accessed on 19 July 2021).
26. ZhiGang, T.; DongDong, Z.; XiaoJie, Y.; JiaMin, W.; Yu, S. Evaluation of Open-Pit Mine Security Risk Based on FAHP-Extenics Matter-Element Model. *Geotech. Geol. Eng.* **2020**, *38*, 1653–1667. [CrossRef]
27. Bise, C.J. Mining Engineering Analysis. *Min. Eng. Anal.* **1986**. Available online: https://www.osti.gov/biblio/6685713 (accessed on 19 July 2021).
28. Fisonga, M.; Mutambo, V. Optimization of the fleet per shovel productivity in surface mining: Case study of Chilanga Cement, Lusaka Zambia. *Cogent Eng.* **2017**, *4*. [CrossRef]
29. Nguyen, H. Determination of Shovel—Truck Productivities in Open—Pit Mines. 2019. Available online: https://www.researchgate.net/publication/308305875_DETERMINATION_OF_SHOVEL_-TRUCK_PRODUCTIVITIES_IN_OPEN_-PIT_MINES (accessed on 19 July 2021).
30. Herrera Herbert, J. Seguridad, Salud y Prevenció n de Riesgos en Minería. Bachelor's Thesis, Universidad Politécnica de Madrid, Madrid, Spain, 2008.
31. Stemn, E.; Bofinger, C.; Cliff, D.; Hassall, M.E. Examining the relationship between safety culture maturity and safety performance of the mining industry. *Saf. Sci.* **2019**, *113*, 345–355. [CrossRef]
32. Xiao, D.; Shan, F.; Li, Z.; Le, B.T.; Liu, X.; Li, X. A target detection model based on improved tiny-yolov3 under the environment of mining truck. *IEEE Access* **2019**, *7*, 123757–123764. [CrossRef]
33. Rojas Ramos, F. Detección de Objetos Usando Cámaras y Sensores LIDAR. Bachelor's Thesis, Benemérita Universidad Autónoma de Puebla, Puebla, Mexico, 2020.
34. Adams, K.K.; Bansah, K.K. Review of Operational Delays in Shovel-Truck System of Surface Mining Operations. In Proceedings of the 4th UMaT International Mining and Mineral Conference, Tarkwah, Ghana, 3–6 August 2016; pp. 60–65.
35. Editec, G.E. Detección de Proximidad para Equipos Mineros en Operaciones de Extracción. Available online: https://www.mch.cl/reportajes/deteccion-de-proximidad-para-equipos-mineros-en-operaciones-de-extraccion/ (accessed on 24 September 2020).
36. Candia, R.C.; Hennies, W.T.; Azevedo, R.C.; Soto, I.G.A.J.F. Análisis de accidentes fatales en la industria minera peruana. *Boletín Geológico y Minero de España* **2010**, *121*, 57–68.
37. Fiscor, S. Productivity Considerations for Shovels and Excavators. *Eng. Min. J.* **2007**, *208*, 38–42.

mining

MDPI

Article

Safe Mining Assessment of Artisanal Barite Mining Activities in Nigeria

David Oluwasegun Afolayan [1,2,*], Azikiwe Peter Onwualu [1], Carrick McAfee Eggleston [2], Adelana Rasak Adetunji [1,3], Mingjiang Tao [2] and Richard Kwasi Amankwah [1,4]

1 Department of Material Science and Engineering, African University of Science and Technology, P.M.B. 681, Abuja 900109, Nigeria; aonwualu@aust.edu.ng (A.P.O.); aderade2004@yahoo.com (A.R.A.); rkamankwah@umat.edu.gh (R.K.A.)

2 Department of Civil and Environmental Engineering, Worcester Polytechnic Institute, 100 Institute Road, Worcester, MA 01609, USA; ceggleston@wpi.edu (C.M.E.); taomj@wpi.edu (M.T.)

3 Department of Material Science, Obafemi Awolowo University, Ile-Ife 220282, Nigeria

4 Department of Mineral Engineering, University of Mines and Technology, Tarkwa P.O. Box 237, Ghana

* Correspondence: dafolayan@aust.edu.ng; Tel.: +1-508-410-3977

Abstract: Barite, used in mud formulation, is mined in several places to support the industry. However, there is insufficient literature on the downside of mining and associated hazards, especially in the artisanal barite mining sector. This paper contains three parts. The initial section reviews major causes of mining accidents and health hazards in Nigeria. The second section examines existing but weak institutional frameworks and policies for artisanal and small-scale mining (ASM) in Nigeria. In the third part, data from questionnaires and heavy metal contamination assessment are compared with health and environmental standards to identify and characterize hazards. It was observed that 54% had health challenges traceable to illicit drugs, and 54% were ignorant about the use of safety kits. The UV-Vis, AAS, and ICP-MS analyses confirmed lead, barium, zinc, copper, and iron in the water samples. Index of geoaccumulation (Igeo) and contamination factor (CF) show that water samples are moderate to highly polluted by Pb^{2+}, Ba^{2+}, and highly contaminated. The chronic daily intake assessment and health quotient analysis revealed that the accumulation of lead and barium is possible and can initiate chronic diseases in humans over a long time. Certain safe mining protocols and controls are recommended.

Keywords: mining hazard; safe mining; miners; artisanal barite mining; mining sites

Citation: Afolayan, D.O.; Onwualu, A.P.; Eggleston, C.M.; Adetunji, A.R.; Tao, M.; Amankwah, R.K. Safe Mining Assessment of Artisanal Barite Mining Activities in Nigeria. *Mining* **2021**, *1*, 224–240. https://doi.org/10.3390/mining1020015

Academic Editors: Juan M Menéndez-Aguado, Oscar Jaime Restrepo Baena and Jessica M. Smith

Received: 24 August 2021
Accepted: 14 September 2021
Published: 18 September 2021

Publisher's Note: MDPI stays neutral with regard to jurisdictional claims in published maps and institutional affiliations.

1. Introduction

Mining is one of the world's most dangerous occupations [1]. Over the years, many mining-associated accidents have occurred in various parts of the world, often with significant loss of life [1–10]. Such mining accidents remind us of how dangerous mining jobs can be and how explicitly hazardous underground mining continues to be [11,12]. Similarly, surface mining blasting-related risks (although not specific to underground mining operations) and their consequences could be worsened and may result in mass widespread accidents [13–15].

Mining accidents and fatalities among the Artisanal and Small-scale miners (ASMs) occur in the process of mining metals, minerals, and energy materials (i.e., not construction materials), as shown in Table 1. Thousands of miners die from these mining accidents each year, especially in coal and hard rock mining [16]. Although surface mining is usually less hazardous than underground mining [2,17,18], the participation of artisanal and small-scale miners in barite mining fields has increased the number of mining fatalities across the upper and middle Benue Trough. Artisanal and small-scale mining (ASM) in Nigeria employed about 0.5 million as of 2015 [19], and in 2021 over 2 million. These miners' and mining communities' contribution to societal development is vital. Both occupational and environmental health and safety issues must be addressed at the mines and workplaces objectively.

Table 1. Some cases of mining hazards in Nigeria.

Case Study	Damages/Sources/Causes	Remedy	References
Concentration of ^{226}Ra, ^{232}Th, and ^{40}K in mining dumps.	radiological hazards, high lifetime cancer risk index	No emerging medical health issues were observed. Regular medical Check-up of miners was recommended for early detection and treatment of potential health hazards	[20]
Concentration of Tl, K, Ca, Na and Mg in Au, Pb, and Zn mines' tailings.	High contamination index of Thallium, high ecological and health risks.	Remediation method was recommended, awareness creation	[21]
Concentration of 40K, 238U, and 232Th in tailings from granite mine.	Radiological hazard is within the permissible limit based on UNSCAR	Bioaccumulation/transfer factor level to be monitored	[22]
Concentration of air-borne lead and respirable silica from dry lead ore grinding and processing	high risk of lead poisoning, silicosis and tuberculosis	Wet spray misting used to reduce the mean airborne Pb and respirable silica	[23–25]
Concentration of Cu, Cr, Pb, Cd and Zn in iron ore tailings	serious non-carcinogenic health risk in children, high carcinogenic risk in adults.	research-industry- miners nexus was advocated	[26]
Concentration of As, Sn, Nb, Ta and Cd in surface water and mine tailings (alluvial) soil	moderate arsenic and cadmium Contamination and Geo-accumulation index (CI & GAI)	enforcement of environmental and mining laws to control pollution	

Sources: [20,22–27].

Heavy metal contamination due to mining and mineral processing (washing) has become one of the most silent but significant environmental side effects [28,29]. Studies in the literature have reported on acidification and acid mine drainage associated with the mining of coal, gold, and other minerals containing pyrite and galena (FeS_2 and PbS) [27,30]. Barite is one mineral or ore that has not been examined to pose such a threat [28]. Barite mineral, although non-carcinogenic, may be associated with lead sulphide (PbS) and encrusted with pyrite or iron pyrite microcrystal [31,32]. Sulphuric acid mine runoff is unavoidable when barite tailings containing sulphide minerals are exposed to water and oxygen. The consequence is acidification of water and can increase the release of other heavy metals such as iron, zinc, copper, lead, cadmium, arsenic, and barium.

Previous reviews on safety and risk analysis have shown the relevance of workplace safety models in the safety-critical assessment of risks, either at mines or in any other activities where dangerous tools are used. Several safe assessment methods have been developed to address the quality and productivity of workers that sustain severe accidents at work and uncovered the adverse effect of heavy metal contaminants and other critical environmental threats to human health [33–37]. Researchers have examined ways to domesticate some of these advanced safe mining methods in Nigeria but with little positive results [36,38]. This is because many local miners believed the "advance" safe mining strategies have no direct correlation and cannot provide solutions to the type of mining hazard peculiar to them [19,39]. Moreover, nothing much seems to have changed regarding miners' and government attitudes to mineral exploration. Miners appear to have nothing to worry about despite the dozens of unreported cases of mining accidents. The significance of wearing safety kits such as mining boots, hand gloves, eye goggles, and clothes specifically designated for mining only at the site should be communicated again. There is also a claim that the institutional policy guilds' activities of artisanal and small-scale miners (ASM) caters to chemical contamination due to barite mining. However, the miners' and mining sites' managers are unaware of the safety data sheet, which is a minimum requirement for the operation of mines. Therefore, it is helpful to engage these local miners in discussing prevalent mining accidents and fatalities that have profound health implications and develop safe assessment methods, processes, and programs to prevent the reoccurrence of mining hazards.

This paper reviews mining activities by the artisanal and small-scale miners in Nigeria and presents safe mining strategies. It identifies mining accidents that are peculiar to artisanal and small-scale miners (ASMs), revises existing but weak and inadequate mining policy, and assesses potential mining risks to human health due to mining and social lifestyles of the miners. Questionnaires were administered to local miners (part-time and full-time) within the middle Benue Trough of Nigeria to identify hazards. Water from barite ponds and effluents was also analyzed to characterize associated risks and recommend safe mining protocols and controls, especially for the barite mining sector. Two research questions were investigated in the study. These are: (1) Certain mining accidents and their adverse effect on miners are traceable to miners' refusal to use safe mining kits and (2) Artisanal barite mining contributes to severe heavy metal contamination. Field survey and heavy metal contamination assessment of water in barite ponds and recycled wastewater at barite mine sites validated the research questions.

2. A Review of Status of Artisanal and Small-Scale Mining (ASM) and Safe Mining Practices in Nigeria

2.1. Legal, Regulatory, and Institutional Frameworks of Artisanal and Small-Scale Mining

There are legal and regulatory documents and institutions that govern the activities of artisanal and small-scale miners in Nigeria. Figures 1 and 2 show the existing legal, regulatory, and institutional frameworks for Nigeria's mining sector. Aside from the frameworks, policy objectives guide the everyday activities within the mineral value-chain. These objectives include but are not limited to comprehensive actions on the acquisition of rights, mine ownership requirement and restrictions, minerals processing and export, transfer mineral rights, land use, environmental, mineral titles, health and safety, and constitutional law. Despite these frameworks, Nigeria's mining sector is yet to reach its full potential [40–42].

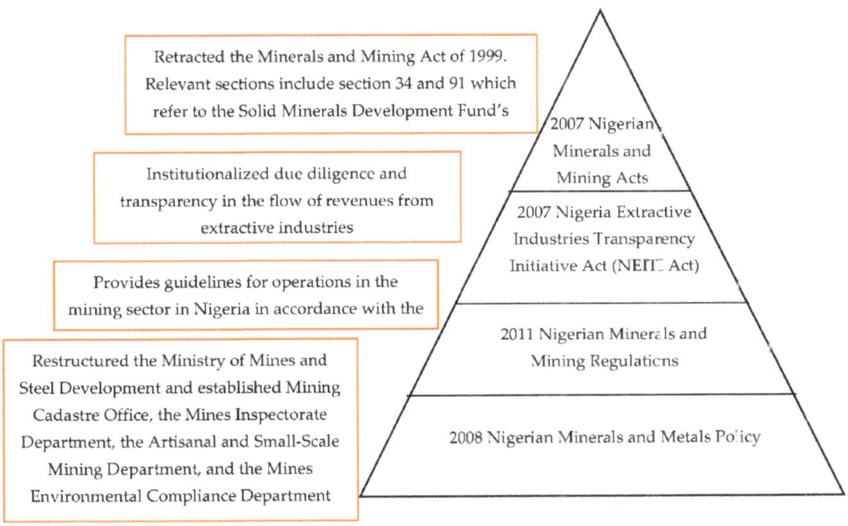

Figure 1. Legal framework for mining in Nigeria (adapted from [40]).

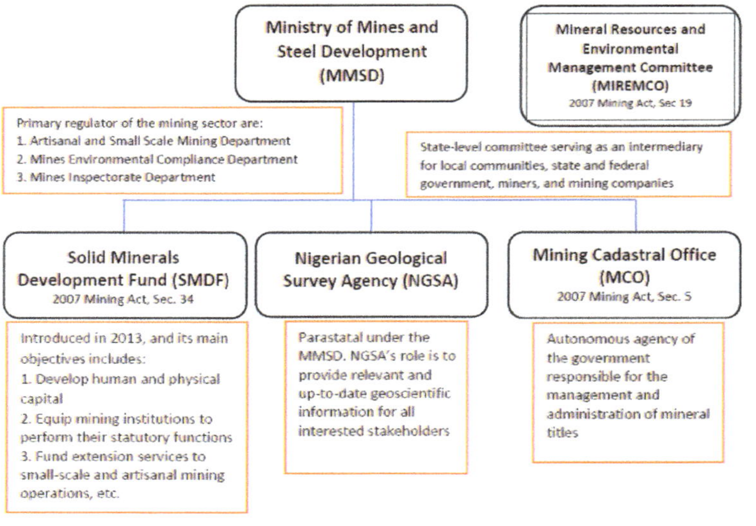

Figure 2. Institutional framework for mining in Nigeria (Modified from [40]).

Research has shown that enacting an Act and introducing laws or policies to drive Nigeria's mining sector can strengthen the regulatory frameworks [40,41,43,44]. However, there were no prior works on health, mine safety, and mining hazard prevention procedures until March 2016, when the Nigerian government acknowledged mercury and lead (Pb) health risks. Mining accidents are not limited to chemical hazards. It also includes every form of harm against the miners, mining communities, and resources located within the mining environment. This set of rules is mandatory and must be enforced by every player within the mining and mineral business [41,42,44].

2.2. Mining Hazards in Nigeria

The sources of hazards associated with the sector include chemical, physical, and mechanical [21,45–53]. Major mining accidents occur due to the use of crude and sharp tools by artisanal and small-scale miners to extract minerals. Some of past and current mining hazards or accidents in different parts of Nigeria are shown in Table 2 and Figure 3. These hazards are traceable to the illegal mining and mineral extraction practices done by artisanal miners in Nigeria. Stone quarrying and solid minerals exploration dominate artisanal and small-scale mining (ASM) activities in Nigeria [45], as shown in Table 2 and Figure 3.

Table 2. Some mining activities and accidents in communities within the Nigerian States.

Mining Hazard/Accidents	Activities/Year	Locations	References
Air pollution (dusts, airborne Si, Ca), eyes damage asthma, damage to farm and cola-nut plantation	limestone quarry, cement production, lead mining (2013 till date)	Shagamu, Ewekoro (Ogun State), kalambana, Wumo, Kwakuti (Sokoto State), Ashaka (Gombe State), Jakura	[21]
Flooding, mysterious death, abandoned mines, contaminated lands, exposure to carcinogenic/radioactive substances.	Tin, columbite and clay mining (1960 till date)	Barkin-Ladi, Bukuru, Bossa, Riyom district (Plateau State).	[27]

Table 2. *Cont.*

Mining Hazard/Accidents	Activities/Year	Locations	References
Heavy metals water contamination, damaged ecosystem. airborne silica, land degradation	Coal, gold, and sand mining (2010–2013)	Enugu, Igun-Ijesha (Oguin State), Efikpo (Ebonyi State), Abeokuta, Owode, Ifo, Ado-Odo, Ofa, Ewekoro, Shagamu (Ogun State), Lagos State	[54]
Water and land degradation, pollutions	Marble mining (2010–2014)	Azara, Wuzue, Benu, Uywa, Lafia (Nassarawa) Luku, Minna (Niger State), Onyeama (Enugu State)	[54]
Death, mine collapse	Gold mining	Zawan (Plateau State)	

Sources: [23,24,54,55].

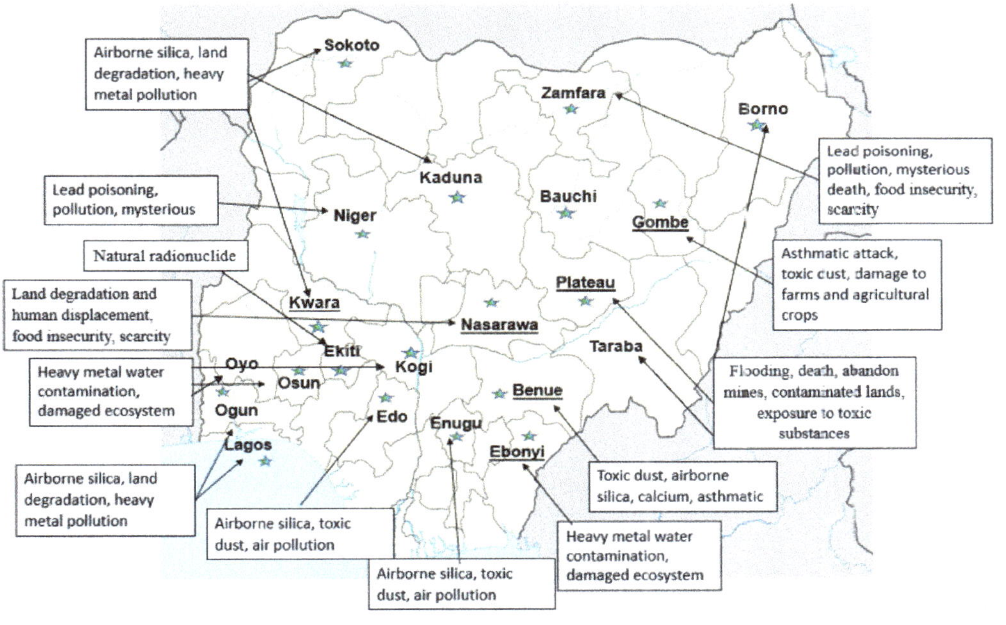

Figure 3. Mining Hazard Map of Nigeria [24,54,56].

2.3. *Safe Mining Methods for Local and Global Mining: Precautions and Control Measures*

Within the last 25 years, there have been increased safety regulations, safer machinery development, training, and education initiatives for miners in Nigeria and Africa in general. However, this has not changed the fact that mining is still a dangerous profession [1]. Before discussing potential accidents and risks in mining, it is vital to consider the average miner work shift based on human resource management. Typically, miners work in a 12 h shift at the underground mine while others work throughout the whole week or remain at a mining camp for months before returning home [17]. Miners are expected to be physically, mentally, and psychologically sound and healthy to achieve overall safety in mines. Strict adherence to safety procedures such as the use of respirators, ventilation systems, and ear protectors will go a long way to reduce cases of mining accidents, injuries, and fatalities [2]. Some of the safety practices and challenges include those involving behavioural guidelines, communication, vehicle interactions, explosives, and the role of enforcement agencies [2,17,57–62].

3. Materials and Methods

3.1. Survey of Miners

The state of hazards within the barite mining industry was examined using surveying tools. Thirty-eight (38) unstructured questionnaires were distributed to miners who specialize in barite mining. Twenty-seven (27) out of thirty-five (~35) barite miners in the community completed and returned the questionnaires. The questionnaire was designed strictly as safety information-seeking procedures based on the major objective of the safety training. The questionnaire also serves as a pre-training/pre-workshop tool or quiz used to identify and assess miners' health concerns, and to develop training manual(s)/choose efficient communication method(s) that address the peculiar needs of the miners under the study.

Approval for the research was obtained from relevant authorities. No medical procedures were observed, as no human body fluids or organs were used for any form of analysis or medical tests. The survey examines why miners refused to use mining boots, gloves, goggles, and clothes contained in the safety mining kits. The entire study attempts to assess and characterize potential health hazards caused by artisanal and small-scale mining (ASM) activities. Questions were read to miners who could not read.

3.2. Chemical Analysis and Risk Assessment

Quantitative risk assessment and health hazard analysis were done in accordance with environmental standards and procedures. Water samples were collected from abandoned barite ponds and wastewater from barite washing and stored in polyethylene bottle (PET) at room temperature. Two ml of the water samples were measured into the cuvette and filled to a mark. The dissolved elements in water samples such as Pb^{2+}, Ba^{2+}, Zn^{2+}, Fe^{2+}, and Cu^{2+} were analyzed colorimetrically using a Shimadzu UV-1900 UV-Vis Spectrophotometer. Tailings effluents was prepared in accordance to standards reported in [63]. The metallic content in the water samples were analyzed using atomic absorption spectrophotometer (AAS), Model: A-Analys 100. The liquid-liquid extraction method (LLEM) was employed in the absorption or digestion of the sample [64]. The elemental composition of the samples was measured using PerkinElmer ICP mass spectrometer, NexION™ 350X. The digestates of barite tailings or extracts were diluted to 1% (100 times).

The index of geoaccumulation (Igeo), contamination factor (CF), chronic daily intake (CDI), and health risk (HQ & HI) are computed for Pb^{2+}, Ba^{2+}, Zn^{2+}, Fe^{2+}, and Cu^{2+} using the data from the USEPA (United States Environmental Protection Agency) and DEA (South Africa Department of Environmental Affairs). Igeo, CDI, CF, and HQ, were computed according to procedures reported in the literature. Each parameter was calculated using Equations (1)–(4) [48,56,65–71]

$$Igeo = log_2 \frac{C_n}{1.5 \times B_n} \tag{1}$$

$$Contamination\ factor = \frac{Mean\ metal\ concentration}{Concentration\ of\ elements\ in\ background\ sample} \tag{2}$$

$$CDI \left(\frac{\mu g}{kg\ day} \right) = \frac{C_{MW} \times I_R}{B_W} \tag{3}$$

$$HQ = \frac{CDI_{non\ carcinogenic}}{RfD} \tag{4}$$

where C_n is the concentration of metal in water samples, B_n is the metal concentration in water before the introduction of metals due to mining activities, C_{MW} is the concentration of heavy metals in water, B_W, and I_R are the body weight and daily water ingestion rate, *HQ* is hazard quotient, *RfD*: reference dose factor, NOAEL: No-Observable Adverse effect level.

4. Results

4.1. Characteristics of Survey Respondents

Figure 4 revealed the level of awareness of mineworkers about the minimum safety required during the mining operations. More than 92% of the miners surveyed were male, and ~64% of the miners who answered the survey were above 25 years-old. It was clear that most miners are young adults, and over 50% of the barite artisanal miners in the study have only basic school education or had no formal education. The miners' biodata showed that many miners only understand local languages and may need to be trained on safe mining methods using local language to communicate essential details.

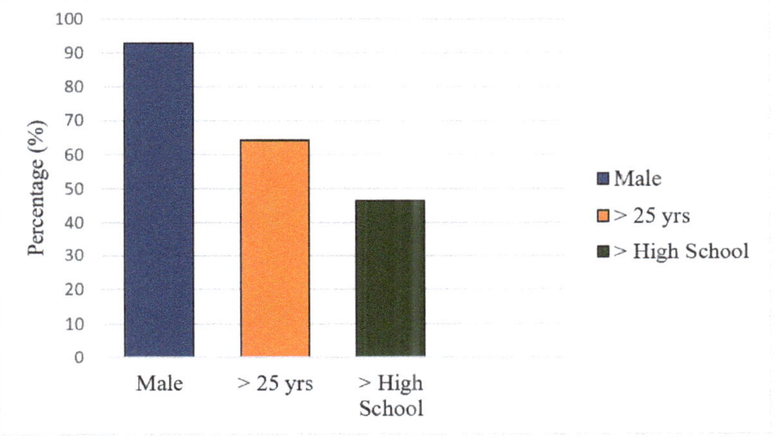

Figure 4. Characteristics of survey respondents (barite miners) showing human participation and performance at the barite mining site.

4.2. Health Hazards of Miners

Part of the survey sought to know the previous and present health challenges of miners within the barite field under the study. Figure 5 shows that ~54% of the miners that responded to the survey agreed they have health challenges traceable to illicit drug intake such as stimulants; 17.9% of the respondents had experienced specific symptoms such as headache, stomach-ache, body weakness, and difficulty breathing. Such health issues may be traceable to rigorous mining activities and exposure to poisonous substances [72,73]. In comparison, 28.7% argued that they do not have any health issues. Also, 53.6% of the miners were ignorant of the benefits of using safety kits for mining, while 46.4% of the miners use safety kits but not at all times. Mine workers were exposed to certain risks, either knowingly or ignorantly, and become most vulnerable to sickness, air-borne diseases, and perhaps death because of insufficient knowledge about the risks associated with the mining profession.

Miners are subjected to long-time exposure to heavy metals contamination. Water used for washing minerals accumulates in ponds near the mining sites and are used for domestic purpose. Potential oral and dermal ingestion are assessed by analyzing water from barite ponds and tailings.

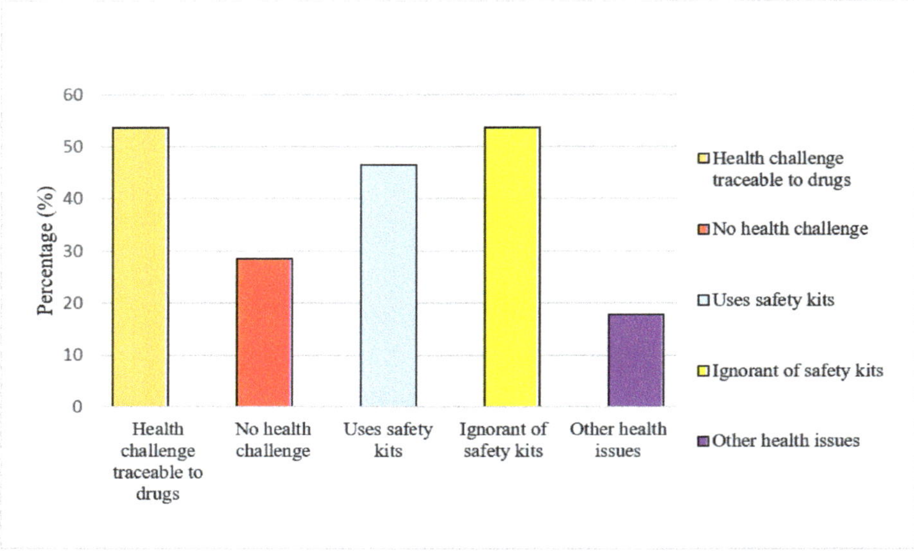

Figure 5. Health and safety issues in the mining sites under study.

The ultraviolent-visible (UV-Vis) spectra in Figure 6 identify absorbance bands showing the weak d-d transition of some identified transition metal complexes in the water samples. This indicates the formation of complexes of the transition metals in the octahedral fields as the d-orbital splits. The calibrated UV-visible spectrophotometer signifies and matches the transition metals in solution using the colour of the d-block compound. The peak absorbance wavelength of 675 nm is assigned to Cu^{2+}, and the visible absorbance band that stretches from 960–980 nm indicates electronic excitations for Fe^{2+}, V^{4+}, and Ni^{2+}. Similarly, the atomic absorption spectroscopy (AAS) identifies and measures the concentration of Zn^{2+}, Pb^{2+}, Cd^{2+}, Fe^{2+}, and Cu^{2+}, as shown in Table 3. The result indicates that Zn^{2+}, Pb^{2+}, Cd^{2+}, Fe^{2+}, and Cu^{2+} as transition metal ions may be present in the water samples associated with the mining site, as indicated by the barite tailings. However, the concentration of copper and cadmium available in the site is less when compared with the World Health Organization (WHO) Standards or limits. The available concentration of lead and iron were 113.8 mg/kg and 15.6 mg/kg, respectively. In contrast, the WHO limits for these elements are pretty small, as shown in Table 3. Fe, Pb, and Cu exceed the WHO allowable limit and remain a potential threat to the mine workers and the host community.

Table 3. AAS analysis water sample TB (completely leached tailings) showing the concentration of heavy metals at barite mining sites in the Middle Benue Trough, Nigeria (Results were compared to WHO data in [3,17]).

Heavy Metals	Proportion	
	Barite Mining Site (mg/L)	WHO Allowable Limit (mg/L)
Zinc	3.905	3.000
Iron	15.6094	0.300
Copper	0.3024	2.000
Lead	113.8127	0.010
Cadmium	0.0008	0.0030

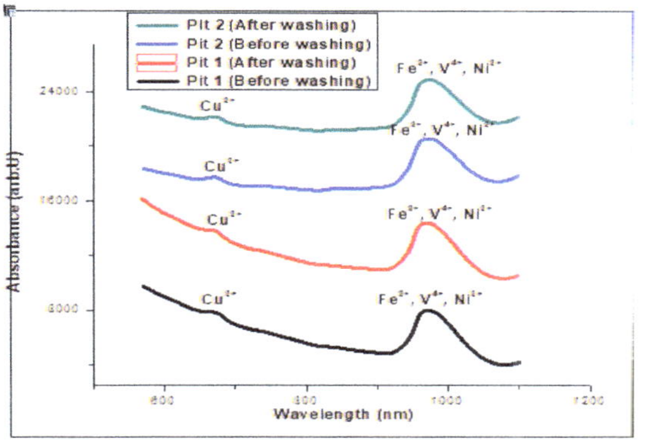

Figure 6. UV-Vis spectrograph for the elemental composition of water from the mined pits (UV-visible spectra of transition metals complexes identified in the water samples showing weak d-d absorbance bands at 675 nm and is assigned to Cu^{2+}, and absorption bands that stretch from 960 to 980 nm posted to Fe^{2+}, V^{4+}, and Ni^{2+}, respectively).

The inductively coupled plasma mass spectroscopy (ICP-MS) results in Figure 7 shows that zinc, copper, and cadmium are below the maximum allowable limit set by World Health Organization (WHO), European Union (EU), Nigerian Industrial Standards (NIS), United States Environmental Protection Agency (USEPA), and China Ministry of Health National Standards (CMHNS) for ecological and health safety. However, the content of Fe in TB1 is relatively higher than the maximum allowable limits set by the governing standards. Similarly, Pb in TB1 is above the health and environmental risk levels recommended by the local and international agencies. This outcome indicates that the water used in the ore washing will result in water pollution and heavy metals' ingestion if returned to rivers and streams used by people. On the contrary, there was no evidence of cadmium contamination in the digestates of mine tailings or tailing effluents in the current study, as shown in Figure 7.

Contamination assessment of mine water samples in Table 4 shows that the index of geoaccumulation (Igeo) for Ba, Cu, and Fe in TB1 is between 0 and 1. The barite ponds and rivers are moderately polluted by Ba, Cu, and Fe. Similarly, Igeo of Pb in TB1 is above 6 ($\geq$6). This indicates Pb extremely pollutes the ponds. The contamination factor (CF) of Ba in TB2 and Fe in TB2, Zn, and Cu in both samples are less than 1 (CF < 1). This implies that the water samples are lowly contaminated by Ba, Fe, Zn, and Cu and cannot pose any substantial risk to the health of miners and residents of the mining sites.

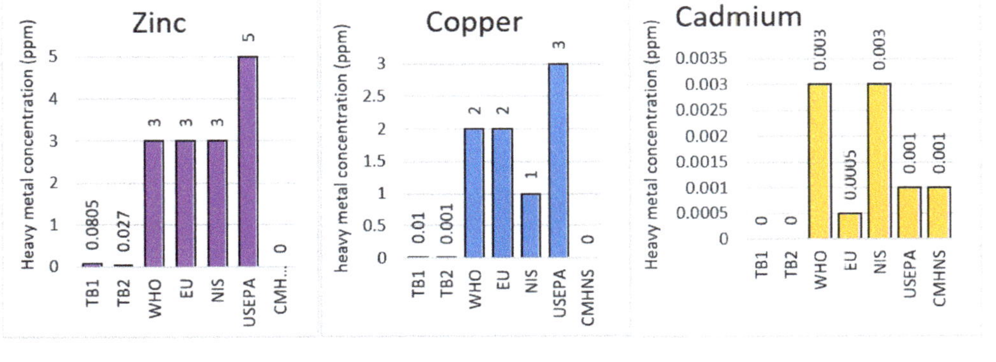

Figure 7. *Cont.*

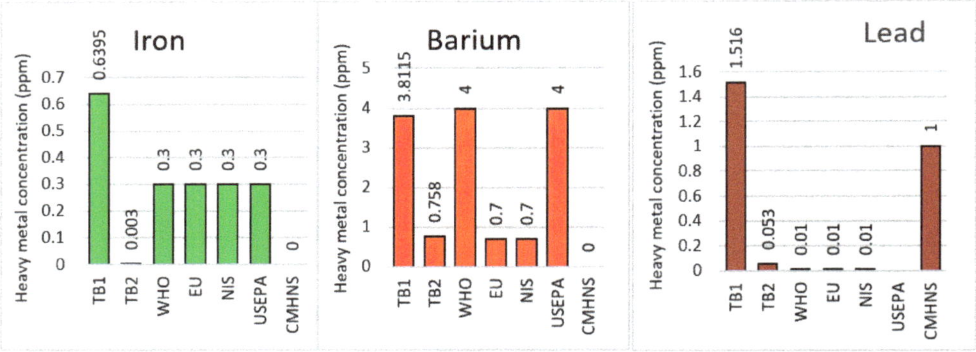

Figure 7. ICPMS Analysis. The concentration of heavy metals associated with artisanal barite mining (ABM) at some mining sites within the Middle Benue Trough, Nigeria. WHO (World Health Organization); EU (European Union); NIS (Nigerian Industrial Standard); USEPA (United States Environmental Protection Agency); CMHNS (China Ministry of Health National Standards).

Table 4. Contamination Assessment of Heavy metals in mines water and tailing effluents.

Elements	Ba	Pb	Zn	Cu	Fe
(Igeo)					
TB1	0.344	7.659	−6.546	0	0.510
TB2	−1.985	2.821	−8.118	−10.966	−7.243
(CF)					
TB1	1.906	303.2	1.61×10^{-2}	9.2×10^{-3}	2.132
TB2	0.379	10.600	5.4×10^{-3}	1.5×10^{-4}	0.010
(CDI) Adult					
TB1	4.47×10^{-2}	1.78×10^{-2}	9.45×10^{-4}		7.50×10^{-3}
TB2	8.90×10^{-3}	6.22×10^{-4}	3.17×10^{-4}	1.17×10^{-5}	3.52×10^{-5}
(CDI) Child					
TB1	4.17×10^{-2}	1.66×10^{-2}	8.82×10^{-4}		7.00×10^{-3}
TB2	8.30×10^{-3}	5.80×10^{-4}	2.95×10^{-4}	1.10×10^{-5}	3.29×10^{-5}

On the other hand, the CF for Ba in TB1 and Fe in TB1 is between 1 and 2.999, and Pb in TB1 also exceeds 6 (>6). Pb moderately contaminates the barite ponds and other water resources. Also, the chronic daily intake (CDI) for Ba, Pb, Zn, Fe, and Cu in barite ponds or mine water and tailing effluents is between 1.17×10^{-5} mg/kg day and 4.47×10^{-2} mg/kg day for an adult, 1.10×10^{-5} mg/kg day, and 4.17×10^{-2} mg/kg day for children. The result presents the possible consequence of long-term exposure to heavy metals and classifies the toxicity level as acute or chronic.

Table 5 indicates that health quotients (HQs) of Zn, Cu, and Fe for the tailings (TB1 & TB2) are less than 0.1. Such HQ is classified as No risk (HQ < 0.1) and cannot lead to adverse health implications in a short time. The presence of Ba and Pb in TB2 poses a relatively low risk to health which shows that some precautionary measures should be taken to avert negative health consequences. However, Pb in TB1 contributes medium to high risk (for 1 < HQ < 4, and HQ > 4). Thus, an adverse effect non-carcinogenic risk is expected. Table 5 also shows that health indexes (HIs) of heavy metals in TB2 for adults and children are below 1. However, HIs for TB1 are greater than 1. For children and adults that drink up to 2 L of water from water sources contaminated by TB1, a cumulative HI of 5.81 indicates elevated non-carcinogenic risks (Table 5).

Table 5. Risk characteristics [Hazard Quotient (HQ) and health index (HI)] of Heavy metals in mines water and tailing effluents.

Elements	Health Quotient (HQ)					Health Index
	Ba	Pb	Zn	Cu	Fe	
(HQ) Adult						
TB1	6.39×10^{-1}	1.27×10^{1}	3.15×10^{-2}		1.07×10^{-2}	1.34×10^{1}
TB2	1.27×10^{-1}	4.45×10^{-1}	1.06×10^{-2}	2.94×10^{-4}	5.03×10^{-5}	5.83×10^{-1}
(HQ) Child						
TB1	5.97×10^{-1}	1.19×10^{1}	1.11×10^{-2}		4.91×10^{-3}	1.25×10^{1}
TB2	1.19×10^{-1}	4.15×10^{-1}	9.86×10^{-3}	2.74×10^{-4}	2.30×10^{-5}	5.44×10^{-1}

5. Discussion

The survey results shown in Figure 4 agree that artisanal barite mining is dominated by men (mostly young adults) and has a lower literacy level as reported on the general status of artisanal and small-scale mining (ASM) in Nigeria. Previous research has shown that artisanal miners of gold, gemstones, diamond, galena, limestone, zinc have similar gender distribution and are exposed to peculiar risks and difficult tasks associated with their profession. Miners are predominantly unskilled and semi-skilled, as observed with artisanal miners that specialize in gold, gemstone, granite, and sand mining. This agrees with the general state of several mining sites managed by artisanal and small-scale miners in Nigeria [19,74–78].

In the current survey, it was quite true that some of the miners felt their present medical conditions are due to factors other than mining, as shown in Figure 5. Several works reported in the literature have shown that all miners are vulnerable to mining hazards aside from previous medical conditions, except for those using complete protective kits during mining [45,73]. Artisanal miners are exposed to dust risk, which lowers the Forced Expiratory Volume (FEV) and Forced Vital Capacity (FVC). Such results have shown that miners that abuse drugs as stimulants may not experience reduced lung function (fibrosis), defective oxygen diffusion, and impaired pulmonary function in the short term. However, exposure to heavy metal contamination would further worsen the present medical conditions [19,45,73,79–81].

Post-survey discussion with miners reveals that artisanal barite miners do not have the financial capacity to fund bills of medical examinations. Most artisanal miners earn lower than the cost of medical treatment. They would prefer self-medication or visit a traditional medical practitioner for medical consultation and treatment as no medical facilities and personnel available. Miners illicitly use nicotine to fight body weakness and other symptoms that requires an adequate medical examination. Also, it is uncertain whether owners of mining sites offer medical care to miners as there is no part of the mining policy or institutional frameworks that compelled or enforced employers to provide for the medical care of miners. Miners are encouraged to use safety kits during mining and seek medical attention when necessary. The need for annual medical outreach to mining sites is recommended for medical counseling, diagnosis, treatments, and referral of miners with severe medical conditions to access medical facilities.

High values of HQs for Ba and Pb increase HI's value for water sample TB1. However, the case is different for sample TB2, posing no observable hazard to human health. The use of such water for various applications and eating aquatic lives such as fishes loaded with heavy metals is unsafe. Also, the heavy metal contamination risk assessment revealed that water from barite ponds and wastewater returned into the river are contaminated by lead and barium. The chronic daily intake (CDI), health quotient (HQ), and health index (HI) for these heavy metals in the water samples suggest that an adverse effect due to non-carcinogenic risk is expected. The use of affordable water filters such as carbon filters specifically designed to remove lead and Ba will help to reduce the quantity of heavy metals consumed in drinking water.

5.1. Major Inhibitors to Safe Mining Methods in Nigeria

The foremen, managers, and owners of mining sites, mineral processors within the mining industries, and academia, as stakeholders, were interviewed verbally to identify major inhibitors to safe mining in Nigeria. The inhibitors identified include funding, lack of enforcement, infrastructural needs, and insecurity.

Project Funding: The Nigerian government has done a lot through the Federal Ministry of Mines and Steel Development (MMSD) in the reform of institutional framework, establishment of ASM Directorate, Solid Minerals Development Fund (SMDF), Mineral Sector Support for Economic Diversification Project. However, some of the stakeholders in the industry and research institutions complained that funds for the projects hardly get to the mine inspectors to develop safety procedures and protocols.

Regulations and Sanctions: Although many regulations and sanctions have been established, implementation has been lacking. Mine inspectors hardly visit mine sites, and minimal awareness is created among the miners on safety and health hazards.

Infrastructural Collapse and Decay: The infrastructural imbalance within the country has completely paralyzed the power, transportation, mines, and minerals sector of the economy. However, the outright privatization of electricity generation and distribution and rail transportation should encourage investments in mining equipment importation for local mineral beneficiation and development of mines.

Security and illegal mining: Most recent and ongoing security challenges within the middle belt, Northeastern and Niger-delta regions of Nigeria can be addressed by developing a robust corporate social responsibility program to alleviate the suffering of the people living within the mineral mining and processing communities. The enactment of the mining act and collaborations among the foreign investors and experts will assist the Nigerian government in the development of a workable mining framework and a road map significantly required for relevance within an acceptable safe mining operation [17,82].

5.2. Impact of COVID-19 on Health of Miners

The first official case of the coronavirus disease 2019 (COVID-19) pandemic was announced in Nigeria on February 27, 2020 [83,84]. In the advent of the COVID-19 pandemic, Nigeria's mining industry experienced sudden downtime, reducing its contribution to the national gross domestic product (GDP). The recent drop-in commercial activities and demand for minerals has also worsened the situation. Also, there are cohorts of individuals facing health and financial challenges during the pandemic. Aside from the older people, miners and mining community' respiratory health is at stake due to the fact that some miners have pre-existing medical complications [73,80,85]. There is, however, no specific data or literature on incidents of COVID-19 related cases or the death of miners. Other subsidiary concerns among the artisanal and small-scale miners, who do not have a stable income for feeding and medical tests, surround the ability to continue routine medical examination and treatment during the pandemic. Therefore, the participation of private health providers and global aid agencies is critical at this point.

In the real sense, the right time to implement innovative and strategic plans, cultivate safety information-seeking behavior in artisanal and small-scale miners (ASMs), and enforce safe mining practices to ensure that miners and the mining activities are safe, is now. Such plans are not limited to remote collaboration, adoption of digital capabilities, safety training on the use of safe mining kits, strict observance of work ethics, occupational and environmental health safety protocols, and personal hygiene in addition to local CDC protocols on COVID-19 prevention, and vaccination of miners. Also, in collaboration with the Capstone team in the United Kingdom, the Nigerian government is reassessing the existing roadmap for mineral exploration amidst new challenges and opportunities due to the pandemic [40,86,87].

5.3. Policy Imperatives and Strategies for Fostering Safe Mining

Mining in Nigeria is regulated by the Constitution of the Federal Republic of Nigeria, 1999, the Nigerian Minerals and Mining Acts, 2007. The Nigerian Minerals and Mining Regulations, 2011 are the significant regulations and policies that control the artisanal and small-scale mining (ASM) activities in Nigeria. These policies directly address issues related to mineral exploration, environmental protection, and safety [19,88–90]. Policies on the environment, health, and safety have been the focus of this study. Although laws should enforce strict observance of these policies for all miners, only legal holders of mineral titles can be tracked. There are reports on Nigeria's government effort to formalize over 1.5 million artisanal and small-scale miners (ASMs) into cooperative groups [40,89,91]. However, information available to miners is limited.

Mine Inspectors and Mine Cadastral Officer are responsible for information dissimilation, but their ratio to ASMs is about 1:200 to 1:10,000. There is an urgent need to strengthen information aids and sources to formalize artisanal and small-scale miners in Nigeria. An information sharing framework can be supported by government declaration for a Miners' Day, a public holiday entirely given to massive sensitization on safe mining issues, safety education and awareness, medical outreaches, and miners networking. Considering mining as a hazardous endeavor, formalizing ASMs into groups will ensure adequate operations management and encourage the participation of relevant stakeholders such as Medical Doctors and Paramedics, rock mechanics, and mining engineering experts. Given the above, existing policies should guarantee safe mining at all mining sites in Nigeria.

There exists a generalized future mining plan in Nigeria called Nigerian Mining Road Map, but the content only speaks to the public without any commitment to ensure its compliance. As earlier mentioned, owners of mining sites and the government are more concerned with the business of mining and not the quality of mineral extraction, safety of life, and the mining environment. The road map proposes the path to mining prosperity and not to ensuring a responsible and sustainable mineral extraction. However, as part of the plan to diversify the economy due to the pandemic, the Ministry of Mines and Solid Minerals Development (MMSD) is considering using Science and Technology in solid mineral exploitation. This includes the use of satellites for mining data acquisition for solid mineral exploration and Artificial Intelligence (Al) to ensure mining safety and efficiency of mineral processing methods. There is a need to adopt an automated safe mining strategy or incorporate mine-based technology such as mine remoting and an automated mining system. This is key to envisioning sustainable barite mining; however, a stable power supply (electricity) is needed to drive this technology contained in the mining road map.

6. Conclusions

This study identifies and reviews mining accidents peculiar to artisanal and small-scale mining (ASM) to re-iterate that mining accidents have severe consequences on miners and their environment. It revises existing but weak and inadequate mining policy, assessing potential mining risks to human health due to the mining and social lifestyles of the miners. Results show that artisanal miners are exposed to polluted water, air, and farmland. The consumption of water from barite ponds poses a relatively high risk to human health over a long period of time. Therefore, it can be concluded that mineworkers are exposed to a certain level of risks either knowingly or ignorantly due to artisanal barite mining. Adverse non-carcinogenic risks due to Pb and Ba in water and a worsening of health via illicit drug intake are expected. Operational therapy and practices such as sensitization on the danger of drugs to health, the importance of taken sufficient rest, and the use of safety tools and affordable water filter have been recommended to ensure safer artisanal mining activities. To envision the future of barite mining, detailed recommendations on the need for annual medical outreach to mining sites and the use of technology (AI) for future mining were presented. Some peculiar safe mining protocols and controls to reduce the daily chronic intake (CDI) of heavy metals in water (barite pond and tailings) are also mentioned.

Author Contributions: Conceptualization, D.O.A. and A.P.O.; Data curation, D.O.A.; Formal analysis, D.O.A. and A.R.A.; Funding acquisition, D.O.A.; Investigation, D.O.A. and A.R.A.; Methodology, D.O.A.; Project administration, A.P.O.; Supervision, C.M.E. and R.K.A.; Validation, D.O.A.; Writing—original draft, D.O.A.; Writing—review & editing, D.O.A., A.P.O., C.M.E., A.R.A., M.T. and R.K.A. All authors have read and agreed to the published version of the manuscript.

Funding: The study (Ph.D. program for the first author) was funded by Regional Scholarship Innovation Fund (RSIF).

Institutional Review Board Statement: Institutional approval was obtained from the African University of Science and Technology prior to the study.

Informed Consent Statement: Informed consent was obtained from all subjects involved in the study.

Data Availability Statement: Raw data from the study is available on request.

Acknowledgments: Authors acknowledge the support of Worcester Polytechnic Institute (WPI), USA, African University of Science and Technology (AUST), Nigeria, and Pan African Material Institute (PAMI), Nigeria.

Conflicts of Interest: Authors declare that they have no competing financial interest or personal relationship that could have appeared to influence any part of this work as reported in the paper.

References

1. Amponsah-Tawiah, K.; Ntow, M.A.O.; Mensah, J. Occupational Health and Safety Management and Turnover Intention in the Ghanaian Mining Sector. *Saf. Health Work* **2016**, *7*, 12–17. [CrossRef] [PubMed]
2. NIOSH. *Injuries, Illnesses, and Hazardous Exposures in the Mining Industry, 1986–1995: A Surveillance Report*; NIOSH: Washington, DC, USA, 2000; pp. 1–141.
3. Donoghue, A.M. Occupational health hazards in mining: An overview. *Occup. Med.* **2004**, *54*, 283–289. [CrossRef] [PubMed]
4. Bringemeier, D. Inrush and mine inundation—A real threat to Australian coal mines. In Proceedings of the International Mine Water Association, Annual Conference, Bunbury, Australia, 30 September–4 October 2012; Volume 30.
5. Walters, D.; Wadsworth, E.J.K.; Johnstone, R.; Quinlan, M. *A Study of the Role of Workers Representatives in Health and Safety Arrangements in Coal Mines in Queensland*; Cardiff University Press: Cardiff, UK, 2014; pp. 1–115.
6. Geng, F.; Saleh, J.H. Challenging the emerging narrative: Critical examination of coalmining safety in China, and recommendations for tackling mining hazards. *Saf. Sci.* **2015**, *75*, 36–48. [CrossRef]
7. Diwe, K.C.; Duru, C.B.; Iwu, A.C.; Merenu, I.A.; Uwakwe, K.A.; Oluoha, U.R.; Ogunniyan, T.B.; Madubueze, U.C.; Ohale, I. Occupational hazards, safety and hygienic practices among timber workers in a South Eastern State, Nigeria. *Occup. Dis. Environ. Med.* **2016**, *4*, 63–71. [CrossRef]
8. Duarte, A.L.; DaBoit, K.; Oliveira, M.L.S.; Teixeira, E.C.; Schneider, I.L.; Silva, L.F.O. Hazardous elements and amorphous nanoparticles in historical estuary coal mining area. *Geosci. Front.* **2019**, *10*, 927–939. [CrossRef]
9. Le Berre, S.; Bretesché, S. Having a high-risk job: Uranium miners' perception of occupational risk in France. *Extr. Ind. Soc.* **2020**, *7*, 568–575. [CrossRef]
10. De, B. Another Two Mine Fatalities Brings 2020 Total to 58 Deaths. 2021. Available online: https://www.miningweekly.com/ (accessed on 9 December 2020).
11. Hopkins, A.; Maslen, S. *Risky Rewards: How Company Bonuses Affect Safety*; Ashgate Publishing, Ltd.: Surrey, UK, 2015.
12. Quinlan, M. *Ten Pathways to Death and Disaster: Learning from Fatal Incidents in Mines and Other High Hazard Workplaces*; Federation Press: Alexandria, Australia, 2014.
13. Mahdevari, S.; Shahriar, K.; Esfahanipour, A. Human health and safety risks management in underground coal mines using fuzzy TOPSIS. *Sci. Total Environ.* **2014**, *488*, 85–99. [CrossRef]
14. Younger, P.L. Environmental impacts of coal mining and associated wastes: A geochemical perspective. *Geol. Soc. Lond. Spec. Publ.* **2004**, *236*, 169–209. [CrossRef]
15. Young, S.; Balluz, L.; Malilay, J. Natural and technologic hazardous material releases during and after natural disasters: A review. *Sci. Total Environ.* **2004**, *322*, 3–20. [CrossRef]
16. Li, W.; Younger, P.L.; Cheng, Y.; Zhang, B.; Zhou, H.; Liu, Q.; Dai, T.; Kong, S.; Jin, K.; Yang, Q. Addressing the CO_2 emissions of the world's largest coal producer and consumer: Lessons from the Haishiwan Coalfield, China. *Energy* **2015**, *80*, 400–413. [CrossRef]
17. Bansah, K.J.; Yalley, A.B.; Dumakor-Dupey, N. The hazardous nature of small scale underground mining in Ghana. *J. Sustain. Min.* **2016**, *15*, 8–25. [CrossRef]

18. Khanzode, V.V.; Maiti, J.; Ray, P.K. A methodology for evaluation and monitoring of recurring hazards in underground coal mining. *Saf. Sci.* **2011**, *49*, 1172–1179. [CrossRef]
19. Oramah, I.T.; Richards, J.P.; Summers, R.; Garvin, T.; McGee, T. Artisanal and small-scale mining in Nigeria: Experiences from Niger, Nasarawa and Plateau states. *Extr. Ind. Soc.* **2015**, *2*, 694–703. [CrossRef]
20. Daburum, N.H.; Songden, S.D.; Mangset, E.W. Assessment of Radiation Dose with Excess Life Cancer Risk of Mining Dumpsites of Wase, Plateau State, Nigeria. *Afr. J. Med. Phy. Biomed. Eng. Sci* **2019**, *6*, 47–54.
21. JohnPaul, A.A.; Ayodeji, L.T.; Tangfu, X.; Ning, Z.; Liu, Y. Toxicity, uptake, potential ecological and health risks of Thallium (Tl) in environmental media around selected artisanal mining sites in Nigeria. *Int. J. Environ. Anal. Chem.* **2020**. [CrossRef]
22. Orosun, M.M.; Usikalu, M.R.; Oyewumi, K.J.; Achuka, J.A. Radioactivity levels and transfer factor for granite mining field in Asa, North-central Nigeria. *Heliyon* **2020**, *6*, e04240. [CrossRef]
23. Gottesfeld, P.; Tirima, S.; Anka, S.M.; Fotso, A.; Nota, M.M. Reducing lead and silica dust exposures in small-scale mining in northern Nigeria. *Ann. Work Expo. Health* **2019**, *63*, 1–8. [CrossRef]
24. Anka, S.A.; Bello, T.S.; Waziri, A.F.; Muhammad, A.S.; Bello, I.; Nasiru, A.M. Environmental effect of lead combination of mining communities in Zamfara State, Nigeria: A review. *J. Biol. Today World* **2020**, *9*, 1–3.
25. UNEP. *Disaster Waste Management Guidelines*; Joint UNEP/OCHA Environment Unit: Geneva, Switzerland, 2011.
26. Aluko, T.; Njoku, K.; Adesuyi, A.; Akinola, M. Health risk assessment of heavy metals in soil from the iron mines of Itakpe and Agbaja, Kogi State, Nigeria. *Pollution* **2018**, *4*, 527–538.
27. Adewumi, A.J.P.; Laniyan, T.A.; Xiao, T.; Liu, Y.; Ning, Z. Exposure of children to heavy metals from artisanal gold mining in Nigeria: Evidences from bio-monitoring of hairs and nails. *Acta Geochim.* **2020**, *39*, 451–470. [CrossRef]
28. Adamu, C.I.; Nganje, T.N.; Edet, A. Heavy metal contamination and health risk assessment associated with abandoned barite mines in Cross River State, southeastern Nigeria. *Environ. Nanotechnol. Monit. Manag.* **2015**, *3*, 10–21. [CrossRef]
29. Adamu, C.I.; Nganje, T.; Edet, A. Hydrochemical assessment of pond and stream water near abandoned barite mine sites in parts of Oban massif and Mamfe Embayment, Southeastern Nigeria. *Environ. Earth Sci.* **2014**, *71*, 3793–3811. [CrossRef]
30. Laniyan, T.A.; Adewumi, A.J. Evaluation of contamination and ecological risk of heavy metals associated with cement production in Ewekoro, Southwest Nigeria. *J. Health Pollut.* **2020**, *10*, 200306. [CrossRef]
31. Szeleg, E.; Janeczek, J.; Metelski, P. Native selenium as a byproduct of microbial oxidation of distorted pyrite crystals: The first occurrence in the Carpathians. *Geol. Carpathica* **2013**, *64*, 231. [CrossRef]
32. Melekestseva, I.Y.; Tret'yakov, G.A.; Nimis, P.; Yuminov, A.M.; Maslennikov, V.V.; Maslennikova, S.P.; Kotlyarov, V.; Beltenev, V.; Danyushevsky, L.; Large, R. Barite-rich massive sulfides from the Semenov-1 hydrothermal field (Mid-Atlantic Ridge, 13 30.87′ N): Evidence for phase separation and magmatic input. *Mar. Geol.* **2014**, *349*, 37–54. [CrossRef]
33. Dunjó, J.; Fthenakis, V.; Vílchez, J.A.; Arnaldos, J. Hazard and operability (HAZOP) analysis. A literature review. *J. Hazard. Mater.* **2010**, *173*, 19–32. [CrossRef] [PubMed]
34. Wang, B.; Wu, C.; Kang, L.; Reniers, G.; Huang, L. Work safety in China's Thirteenth Five-Year plan period (2016–2020): Current status, new challenges and future tasks. *Saf. Sci.* **2018**, *104*, 164–178. [CrossRef]
35. Wang, B.; Shen, Y.; Saravanan, V.; Luhach, A.K. Workplace safety and risk analysis using Additive Heterogeneous Hybridized Computational Model. *Aggress Violent Behav.* **2021**, 101558. [CrossRef]
36. Qin, Y.; Chen, Z.; Ding, B.; Li, Z. Impact of sand mining on the carbon sequestration and nitrogen removal ability of soil in the riparian area of Lijiang River, China. *Environ. Pollut.* **2020**, *261*, 114220. [CrossRef]
37. Rezaie, B.; Anderson, A. Sustainable resolutions for environmental threat of the acid mine drainage. *Sci. Total Environ.* **2020**, *717*, 137211. [CrossRef]
38. Rezaie, A.; Buresi, M.; Lembo, A.; Lin, H.; McCallum, R.; Rao, S.; Schmulson, M.; Valdovinos, M.; Zakko, S.; Pimentel, M. Hydrogen and methane-based breath testing in gastrointestinal disorders: The North American consensus. *Am. J. Gastroenterol.* **2017**, *112*, 775. [CrossRef]
39. Warra, A.A.; Prasad, M.N.V. *Artisanal and Small-Scale Gold Mining Waste Rehabilitation with Energy Crops and Native Flora-A Case Study from Nigeria*; Elsevier Inc.: Philadelphia, PA, USA, 2018. [CrossRef]
40. Ango, M.; Erdenebat, B.; Tang, K.Y. *Creation of a Sustainable Mining Program through Formalization of Artisanal and Small Scale Miners*; Columbia University Press: New York, NY, USA, 2019.
41. Heffron, R.J. The role of justice in developing critical minerals. *Extr. Ind. Soc.* **2020**, *7*, 855–863. [CrossRef]
42. Macdonald, K.; Lund, M.; Blanchette, M.; Mccullough, C. Regulation of artisanal small scale gold mining (ASGM) in Ghana and Indonesia as currently implemented fails to adequately protect aquatic ecosystems. In Proceedings of the International Mine Water Association Symposium, Xuzhou, China, 18–22 August 2014; pp. 401–405.
43. Akper, P.T.; Ani, L. Legal and Policy Issues in the Development of Nigeria's Mining Sector: Charting the Way Forward. *SSRN* **2020**. [CrossRef]
44. Wireko-Gyebi, R.S.; King, R.S.; Braimah, I.; Lykke, A.M. Local knowledge of risks associated with artisanal small-scale mining in Ghana. *Int. J. Occup. Saf. Ergon.* **2020**, 1–8. [CrossRef]
45. Nwibo, A.N.; Ugwuja, E.I.; Nwambeke, N.O.; Emelumadu, O.F.; Ogbonnaya, L.U. Pulmonary problems among quarry workers of stone crushing industrial site at Umuoghara, Ebonyi State, Nigeria. *IJOEM* **2012**, *3*, 178–185. [PubMed]
46. Taiwo, A.M.; Awomeso, J.A. Assessment of trace metal concentration and health risk of artisanal gold mining activities in Ijeshaland, Osun State Nigeria—Part 1. *J. Geochem. Explor.* **2017**, *177*, 1–10. [CrossRef]

47. Abdullahi, U.; Asuku, A.; Umar, A.; Ahmed, Y.A.; Adam, U.S.; Abdulmalik, N.F.; Yunusa, M.H.; Abubakar, A.R. Assessment of Radon Concentration And Associated Health Implications In Ground Water And Soil Around Riruwai Mine Site, Kano State, Nigeria And Its Environs. *Fudma J. Sci.* **2020**, *4*, 242–246. [CrossRef]
48. Adewumi, A.J.; Laniyan, T.A. Ecological and human health risks associated with metals in water from Anka Artisanal Gold Mining Area, Nigeria. *Hum. Ecol. Risk Assess. Int. J.* **2021**, *27*, 307–326. [CrossRef]
49. Järvholm, B.; Silverman, D. Lung cancer in heavy equipment operators and truck drivers with diesel exhaust exposure in the construction industry. *Occup. Environ. Med.* **2003**, *60*, 516–520. [CrossRef] [PubMed]
50. Njinga, R.L.; Tshivhase, V.M. Major chemical carcinogens in drinking water sources: Health implications due to illegal gold mining activities in Zamfara State-Nigeria. *Expo Health* **2019**, *11*, 47–57. [CrossRef]
51. Article, O. Prevalence of Cough among Adults in An Urban Community in Nigeria. *West Afr. J. Med.* **2011**, *30*, 5.
52. Kyari, F.; Alhassan, M.B.; Abiose, A. Pattern and outcome of paediatric ocular trauma–A 3-year review at National Eye Centre, Kaduna. *Niger. J. Ophthalmol.* **2000**, *8*, 11–16. [CrossRef]
53. Atakpa, A.; Argungu, G.M.; Muawiya, S.; Wase, M.M.; Shuaibu, L.M. Assessment of Mineral Resources in Federal Capital Territory (FCT) Abuja, Nigeria. *Int. J. Res. Appl. Sci. Eng. Technol.* **2019**, *7*, 819–823. [CrossRef]
54. Merem, E.C.; Twumasi, Y.; Wesley, J.; Isokpehi, P.; Shenge, M.; Fageir, S.; Crisler, M.; Romorno, C.; Hines, A.; Hirse, G.; et al. Assessing the ecological effects of mining in West Africa: The case of Nigeria. *Int. J. Min. Eng. Miner. Process.* **2017**, *6*, 1–19. [CrossRef]
55. UNEP/OCHA. *Environment and Humanitarian Action: Increasing Effectiveness, Sustainability and Accountability*; UNEP/OCHA Environment Unit: Geneva, Switzerland, 2014.
56. Muhammad, S.; Tahir Shah, M.; Khan, S. Health risk assessment of heavy metals and their source apportionment in drinking water of Kohistan region, northern Pakistan. *Microchem. J.* **2011**, *98*, 334–343. [CrossRef]
57. Dasgupta, A.K.; Harrison, J. Effects of vibration on the hand-arm system of miners in India. *Occup. Med.* **1996**, *46*, 71–78. [CrossRef]
58. Denton, S.; Allsop, A.; Davies, M.; Al, E. *The Prevention and Control of Fire and Explosion in Mines*; Health & Safety Executive: Geneva, Switzerland, 2001.
59. Frank, T.; Bise, C.J.; Michael, K. A hearing conservation program for coal miners. *Occup. Health Saf.* **2003**, *72*, 106. [PubMed]
60. Manley, L. Should States Serve as Laboratories for Mine Safety Regulation. *Ariz St. LJ* **2009**, *41*, 379.
61. Orogbu, L.O.; Onyeizugbe, C.U.; Chukwuma, E. Safety practice and employee productivity in selected mining firms in Ebonyi State, Nigeria. *J. Res. Bus. Econ. Manag.* **2018**, *10*, 1964–1970.
62. Zhong, B.; Pan, X.; Love, P.E.D.; Sun, J.; Tao, C. Hazard analysis: A deep learning and text mining framework for accident prevention. *Adv. Eng. Inform.* **2020**, *46*, 101152. [CrossRef]
63. Akpan, I.O. Effect of sample treatment on trace metal determination of Nigerian crude oils by Atomic Absorption Spectroscopy (AAS) Technique. *Afr. J. Environ. Pollut. Health* **2005**, *4*, 1–5.
64. Ibe, K.A.; Ogeleka, D.F.; Ani, I.C.; Uyebi, G.O. Suitability of Nigerian barite as a weighting agent in drilling mud. *Int. J. Min. Miner. Eng.* **2016**, *7*, 51–63. [CrossRef]
65. Muller, G. Index of geoaccumulation in sediments of the Rhine River. *Geo. J.* **1969**, *2*, 108–118.
66. Hakanson, L. An ecological risk index for aquatic pollution control. A sedimentological approach. *Water Res.* **1980**, *14*, 975–1001. [CrossRef]
67. EPA; ABD. *Risk Assessment Guidance for Superfund*; Volume I: Human health evaluation manual (Part A); EPA/540/1-89/002; Office of Emergency and Remedial Response, EPA: Washington, DC, USA, 1989.
68. Camp Dresser & McKee, Inc. *Guidelines for Water Reuse, US Environmental Protection Agency*; EPA/625/R-04/108; USEPA: Washington, DC, USA, 2004.
69. EPA, US. *Fact Sheet PFOA & PFOS Drinking Water Health Advisories*; EPA 800-F-16-003; US Environmental Protection Agency: Washington, DC, USA, 2016.
70. Shah, I.; Khan, T.; Hanif, M.; Shah, A.; Siddiqui, S.; Khattak, S.A. Environmental aspects of selected heavy and trace elements of Cherat Coal deposits. *J. Himal. Earth Sci.* **2016**, *49*, 77–85.
71. Adewumi, A.J.; Laniyan, T.A. Contamination, sources and risk assessments of metals in media from Anka artisanal gold mining area, Northwest Nigeria. *Sci. Total Environ.* **2020**, *718*, 137235. [CrossRef]
72. Hessel, G.K.; Sluis-Cremer, P.A. Hearing loss in white South African goldminers. *S. Afr. Med. J.* **1987**, *71*, 354–367.
73. Babatunde, O.; Ayodele, L.; Elegbede, O.; Babatunde, O.; Ojo, O.; Alawode, D.; Ademol, O.; Oluwasey, A. Practice of occupational safety among artisanal miners in a rural community in Southwest Nigeria. *Int. J. Sci. Environ. Technol.* **2013**, *2*, 622–633.
74. WHO. *Environmental and Occupational Health Hazards Associated with Artisanal and Small-Scale Gold Mining*; World Health Organization HQ: Geneva, Switzerland, 2016.
75. Tirima, S.; Bartrem, C.; von Lindern, I.; von Braun, M.; Lind, D.; Anka, S.M.; Abdullahi, A. Environmental remediation to address childhood lead poisoning epidemic due to artisanal gold mining in Zamfara, Nigeria. *Environ. Health Perspect.* **2016**, *124*, 1471–1478. [CrossRef]
76. Nwagwu, W.E. Job Satisfaction of Information Technology Artisans in Nigeria. *Mousaion* **2018**, *36*, 24.
77. Melodi, M.M.; Opafunso, Z.O. An Assessment of Existing Production and Revenue Capacities for Artisanal and Small-Scale Granite Mining in Southwest Nigeria. *J. Min. World Express* **2014**, *3*, 33–37. [CrossRef]

78. Ahmed, Y.M.; Oruonye, E.D. Socioeconomic impact of artisanal and small scale mining on the Mambilla Plateau of Taraba State, Nigeria. *World J. Soc. Sci. Res.* **2016**, *3*, 1.
79. Olley, B.O.; Abikoye, G.E. Predicting Intentions and Continuous Cannabis Use among Smokers. In *Perspectives on Drugs, Alcohol and Society in Africa*; Obot, I.S., Abikoye, G.E., Eds.; Centre for Research and Information on Substance Abuse (CRISA): Jos, Nigeria, 2016; Volume 3, pp. 116–122.
80. Azodo, C.C.; Omuemu, V.O. Perception of spirituality, spiritual care, and barriers to the provision of spiritual care among undergraduate nurses in the University of Lagos, Nigeria. *J. Clin. Sci.* **2017**, *14*, 119–125. [CrossRef]
81. Ralph, O.; Gilles, N.; Fon, N.; Luma, H.; Greg, N. Impact of Artisanal Gold Mining on Human Health and the Environment in the Batouri Gold District, East Cameroon. *Acad. J. Interdiscip. Stud.* **2018**, *7*, 25–44. [CrossRef]
82. Esheya, S.E.; Okoye, P.C.U.; Nweze, P.N.J.; Okonkwo, N.A. Socio-Economic Effects of Chemical Pollution on Agricultural Production in Mineral Mining Communities of South-East Nigeria. *Merit Res. J. Bus. Manag.* **2017**, *5*, 63–68.
83. Kalu, B. COVID-19 in Nigeria: A disease of hunger Respiratory health in athletes: Facing the COVID-19 challenge. *Lancet Respir.* **2020**, *8*, 556–557. [CrossRef]
84. Adebowale, A.S.; Fagbamigbe, A.F.; Akinyemi, J.O.; Obisesan, O.K.; Awosanya, E.J.; Afolabi, R.F.; Alarape, S.A.; Obabiyi, S.O. The spread of COVID-19 outbreak in the first 120 days: A comparison between Nigeria and seven other countries. *BMC Public Health* **2021**, *21*, 129. [CrossRef] [PubMed]
85. Laniyan, T.A.; Adewumi, A.J.P. Potential ecological and health risks of toxic metals associated with artisanal mining contamination in Ijero, southwest Nigeria. *J. Environ. Sci. Health Part A* **2020**, *55*, 858–877. [CrossRef]
86. Amzat, J.; Aminu, K.; Kolo, V.I.; Akinyele, A.A. Coronavirus outbreak in Nigeria: Burden and socio-medical response during the first 100 days. *Int. J. Infect Dis.* **2020**, *98*, 218–224. [CrossRef]
87. Nwagwu, W.E.; Igwe, E.G. Safety information-seeking behaviour of artisanal and small-scale miners in selected locations in Nigeria. *Libri* **2015**, *65*, 71–90. [CrossRef]
88. Afolayan, D.O.; Adetunji, A.R.; Peter, A.; Oghenerume, O.; Amankwah, R.K. Characterization of barite reserves in Nigeria for use as weighting agent in drilling fluid. *J. Pet. Explor. Prod.* **2021**, 1–22. [CrossRef]
89. Mallo, S.J. Mitigating the Activities of Artisanal and Small-Scale Miners in Africa: Challenges for Engineering and Technological Institutions. *IJMER* **2012**, *2*, 4714–4725.
90. Mensah, A.K.; Mahiri, I.O.; Owusu, O.; Mireku, O.D.; Wireko, I.; Kissi, E.A. Environmental Impacts of Mining: A Study of Mining Communities in Ghana. *Appl. Ecol. Environ. Sci.* **2015**, *3*, 81–94. [CrossRef]
91. Saleh, J.H.; Cummings, A.M. Safety in the mining industry and the unfinished legacy of mining accidents: Safety levers and defense-in-depth for addressing mining hazards. *Saf. Sci.* **2011**, *49*, 764–777. [CrossRef]

mining

Article

A High-Fidelity Modelling Method for Mine Haul Truck Dumping Process

Aaron Young * and William Pratt Rogers

Department of Mining Engineering, University of Utah, Salt Lake City, UT 84112, USA; pratt.rogers@utah.edu
* Correspondence: aaron.s.young@utah.edu

Abstract: Dumping is one of the main unit operations of mining. Notwithstanding a long history of using large rear dump trucks in mining, little knowledge exists on the cascading behavior of the run-of-mine material during and after dumping. In order to better investigate this behavior, a method for generating high fidelity models (HFMs) of dump profiles was devised and investigated. This method involved using unmanned aerial vehicles with mounted cameras to generate photogrammetric models of dumps. Twenty-eight dump profiles were created from twenty-three drone flights. Their characteristics were presented and summarized. Four types of dump profiles were observed to exist. Factors that influence the determination of these profiles include the location of the truck relative to the dump crest, the movement of the underlying dump material during the dumping process and the differences in the dump profile prior to dumping. The HFMs created in this study could possibly be used for calibrating computer simulations of dumps to better match reality.

Keywords: dumping; digital transformation; high-fidelity modelling

Citation: Young, A.; Rogers, W.P. A High-Fidelity Modelling Method for Mine Haul Truck Dumping Process. *Mining* **2022**, *2*, 86–102. https://doi.org/10.3390/mining2010006

Academic Editors: Juan M Menéndez-Aguado, Oscar Jaime Restrepo Baena and Jessica M. Smith

Received: 17 December 2021
Accepted: 4 February 2022
Published: 11 February 2022

1. Introduction

Mine-to-mill optimization is a longstanding goal of the mining industry [1–4]. This approach focuses on optimizing the entire process around run-of-mine (ROM) material characteristics, rather than optimizing the unit operations to material characteristics individually. The primary motivation for this approach is ensuring that the priority for optimization is given to the most demanding and costly process (grinding) [5]. Despite its success, one difficulty in holistic mine-to-mill optimization stems from a lack of understanding around material behavior during and between unit operations [4].

Unit operations are defined within the context of mining as the basic steps used to produce mineral value from a deposit. They generally fall into either the category of rock fragmentation or materials handling [6]. Materials handling, for surface truck and shovel mines, comprises three steps, known as loading, hauling and dumping. Dumping commonly consists of the haul truck spotting itself into position and dropping ROM material from the back. This material forms a small heap if it is dumped on a flat surface, or cascades along the edge of a dump face if dumped over a developing dump, stockpile or dump/heap leach [7]. While there are minor differences in each of these earthworks, for the simplification of this article, they are referred to as rock piles [8].

Notwithstanding a long history of using large rear dump trucks in mining, little knowledge exists on the cascading behavior of ROM material during and after dumping. At least two schools of thought comprise the knowledge that does exist. The first school of thought studies ROM cascading behavior from an external perspective. The second seeks to understand the smorgasbord of characteristic properties that govern this behavior internally. Since this paper focuses primarily on the external viewpoint, only a sample of the literature pertaining to the internal philosophy is presented in order to call attention to some of its challenges. McLemore et al. [8] provide a robust review that includes a trove of references from each school of thought, should readers wish to dive deeper.

Work related to factors that influence ROM material behavior from an internal philosophy are numerous, and a complete review of their literature is not within the scope of this paper. A plethora of material property variables relate to these factors, which include shear strength, bulk density, particle size and shape distribution, cohesive properties, friction angle, moisture content, etc. [8,9]. Several issues confound the ability to fully isolate and understand these variables. First, the size, shape and variability of the ROM material are greatly dependent on blasting, which is not yet a fully understood process [10,11]. Second, the successful testing of material property variables depends heavily on sampling and statistical estimation [12]. Furthermore, dump faces are hazardous, and observation of the dumping process is challenging [13]. Additionally, dozers and other equipment handle the material subsequent to dumping, which compound with the issues mentioned previously [7]. Moreover, some complex engineering issues, such as reclamation and slope stability, are commonly considered to be linked to factors involving these material property variables in ways not fully understood, which increases the debate related to them [13–15]. Mines may also be under the assumption that understanding these factors must originate from a first-principles approach, rather than a data-driven statistical and mathematical approach [16].

Traditional viewpoints and operational strategies within the mining industry generally hinder innovation [17,18]. This might be particularly true for each of the three kinds of rock piles. Stockpiles have been seen as only useful to mines as a buffer against production variability [19–21]. Dump/heap leach design optimization involves many meticulous considerations that take precedence over understanding the minutia of the dumping process, which may be considered to be optimized under simple guidelines [22]. Waste dumps have been reserved for material below the economic cut-off grade and, as a result, have historically been given little attention beyond safety and risk management [23]. While these traditional viewpoints would need to change in order for rock pile innovation to be successful, they need only be challenged in order for research in these areas to be justified.

Literature and news articles provide both cases and causes for some changing perspectives that motivate and support the endeavor of this paper. First, even though the primary function of stockpiles remains the same [24], COVID-19 and global supply chain disruptions have placed increased demand on their use [25–34]. Increased knowledge of stockpile assets as the result of understanding the dumping process might afford mining companies a competitive edge in a post-pandemic world [35–38]. Second, while much of the seminal work in heap leach modelling involved small laboratory column studies operating at the particle scale [39–41], there has been a recent shift towards modelling bulk scale phenomena (inter-/intra-particle diffusion, liquid holdup and hysteresis, gas flow, etc. [42,43]) as well as understanding the factors more closely correlated with the dumping process (stratification, segregation, breakage induced by ore stacking, etc. [44–46]). It is true that a better understanding of the dumping process will likely not lead to changes in the construction of heap leaches. However, it may yield an understanding of the gestalts about the bulk phenomena in existing heap leaches, and aid in the development of models that increase their profitability [47]. Third, what was once considered to be below cut-off grade may later become mineable ore [48–52]. Therefore, a deeper understanding of the dumping process as it relates to waste dumps may prove valuable for operations, where old waste dumps become economically viable [53–56]. Finally, the perfunctory amount of documented work, aimed at capitalizing on the opportunities hypothesized previously, may have less to do with the merit of such hypotheses, and more to do with the mining sector's lag in innovation [17,18,57].

To improve the understanding of end-dumping and rock pile construction, we present a method for creating a high-fidelity model (HFM) of the dumping process through a digital transformation approach. Digital transformation [57,58], employs an external philosophy for understanding ROM behavior, and is a process whereby real-world assets and processes are digitally transformed in ways that add value for decision makers. Innovation [59] and emerging technologies [60] make digital transformation possible. Unmanned aerial vehicles (UAVs), also known as drones, are an emerging technology [61] that allows for the digital transformation of the dumping process through data collection. Rock pile faces span large areas, and the nature of the material requires multiple vantage points for sensors or traditional surveys to be effective. By using photogrammetry, UAVs are capable of capturing multiple angles and covering large difficult areas, such as a dump face [47,62]. Thus, an investigation into whether photogrammetry can create the HFM of an individual truck end dump is of interest to the issue of better understanding the dumping process and rock pile construction.

HFMs act as reference systems to reality [63]. These models are commonly used in engineering to calibrate simulation models [64] where there is a need to rapidly prototype many different design permutations [65], or where measuring the real data being modelled is not practical [66]. HFMs are modular elements, and HFM integration is a modular framework that allows each aspect of the full model to be worked on independently [67]. This modularity means that simulations to match HFMs can be worked on independently from the work used to improve how HFMs match with reality.

Digital transformation techniques are fairly new, and little work has been conducted to digitally transform the process of rock pile construction via haul truck end-dumping. Zahl et al. [9] accurately assert that the formation and shape of mine rock piles are based mainly on topography. While this work is foundational to the engineering of rock pile construction, it offers little for the purposes of rock pile digital transformation. Mclemore et al. [8] provide an extensive review on the construction, the factors influencing the shear strength of soil, characterization, the effects of weathering and the stability of rock piles throughout the world. Their review is more informative than that of Zahl et al.; however, it likewise does not consider a digital approach. Zhao [68] developed a real-time 3D modelling and mapping technique for the stockpiles formed by stacking/reclaiming machines for iron ore. Zhao's work is pioneering in the area of the digital transformation of small, intermediary stockpile construction, but does not consider the large rock fills made by haul truck end-dumping. The authors of this present paper previously illustrated a method for modelling and mapping large heap-filled stockpiles using fleet management system (FMS) data [7]. However, while these data were amicable to modelling, no external data were available for the validation of the model. Zhang and Liu [47] employed UAV aerial photography and performed image analysis to investigate particle size distribution along the face of a dump leach. However, while they demonstrated the ability of UAV aerial photography to capture relevant data, they did not capture the volumes of individual dump profiles that could be used as a baseline to digitally transform the end-dumping rock pile construction process. Servin et al. [4] present a digital transformation technique for holistic mine-to-mill distributed particle simulation where the gaps between unit operations are simulated using data from control systems and sensors. While other frameworks for similar integration have been proposed in the past, the focus Servin et al. place on a unit operation-centered framework is consistent with the aims of this paper and the future direction of research in this area.

The method presented in this work demonstrates how to create 3D HFMs of ROM end-dumping from photogrammetry. The method presented is very similar to that of Zhang and Liu [47]. The resulting HFMs are akin to the model developed by Zhao [68], only they involve end-dumped rock piles as opposed to stacker-made intermediary stockpiles. These HFMs could be potentially suitable for the technique presented by Servin et al. [4] for holistic mine-to-mill optimization, which includes large end-dumped rock piles. They may also be used as a modular test bed for the calibration of future simulations of mine haul truck dumping activity. The end goal of the HFMs this method creates is the parameterization of dumping as a unit operation process. Essentially, while the foundational work of Zahl et al. [9] explains that topography is the main factor influencing rock pile shape, the method presented in this paper may be used to increase our understanding of the expected variability and parameters in such topography common to the end-dumping and rock pile construction process, thereby making these shapes more predictable and amicable to modelling.

2. Materials and Methods

2.1. Material and Haulage Equipement

The material studied was low-grade run-of-mine (ROM) gold ore from a surface mining operation near Perth, WA. No additional characterization of the material is required for presenting the concept of this method, although future simulation and modelling work will require more detailed characterization in order to ensure that models are properly calibrated.

The ore was transported and dumped using CAT 793F haul trucks with a payload of approximately 231 metric tons, a struck capacity of 112.6–151 m^3 and an inside bed width of 7334 mm.

2.2. UAV Flights and Photogrammetry Methodology

Field work and UAV flights were conducted in coordination with the mine's survey crew and through the use of their equipment. A total of 23 flights were recorded over the course of five day-shifts. These flights were performed using a DJI Matrice M200 drone with a Zenmuse X5S camera (DJI-Innovations Company Limited, Shenzhen, China). Table 1 shows details about the camera and flight settings used in these flights.

Table 1. Camera and Flight Details.

Camera Details				Flight Details			
Sensor Type	Sensor Size (mm)	Focal Length (mm)	Image Size (Pixels)	Flight Height (m)	Average Flight Area (m^2)	Average Photos Taken	Flight Style
M4/3 CMOS	12.8 × 8.6	8.6	5472 × 3648	60	74,500	65	Snaking Grid

The methodology for the flights was as follows:

1. An initial flight was performed to create and update the local orthomosaic map (approximately 6 min);
2. Ground control points (GCPs) were marked and surveyed (3 to 7 GCPs were used for each flight);
3. A second flight was performed with an adjusted flight path to include ground control points (approximately 7 min);
4. A Quick Map was created from the second adjusted flight path so that the area of interest (dump area) could be readied, and the working flight path programmed;
5. Once steps 1 through 4 were completed, flights were repeated using the site scan Quick Fly software function in the same area along the same flight path to capture the dump face during different time intervals and thereby capture before and after photos of the dumping process during regular operation (60 m flight height, 6 min flight times).

Once the flights were completed, pictures were uploaded to the 3DR cloud processing system, and photogrammetric point clouds were generated from the flight photos using Pix4D software (mapper 4.1 version, Pix4D SA, Lausanne, Switzerland). Absolute and relative accuracy were calculated for each GCP in the X, Y and Z coordinate directions. These calculations were performed automatically as part of the 3DR cloud processing.

Figure 1 shows a top view (XY plane) of an example flight plan with GCP locations and the corresponding image overlap map generated from the flight. All flight plans were similar to Figure 1, with the only changes being the location of different GCPs. The thick black curved line in Figure 1 represents the upper crest of the dumping area, and it is a close approximation to the actual dump crest. As demonstrated by Figure 1, the dump crest area for each flight was photographed with 5 or more images of overlap.

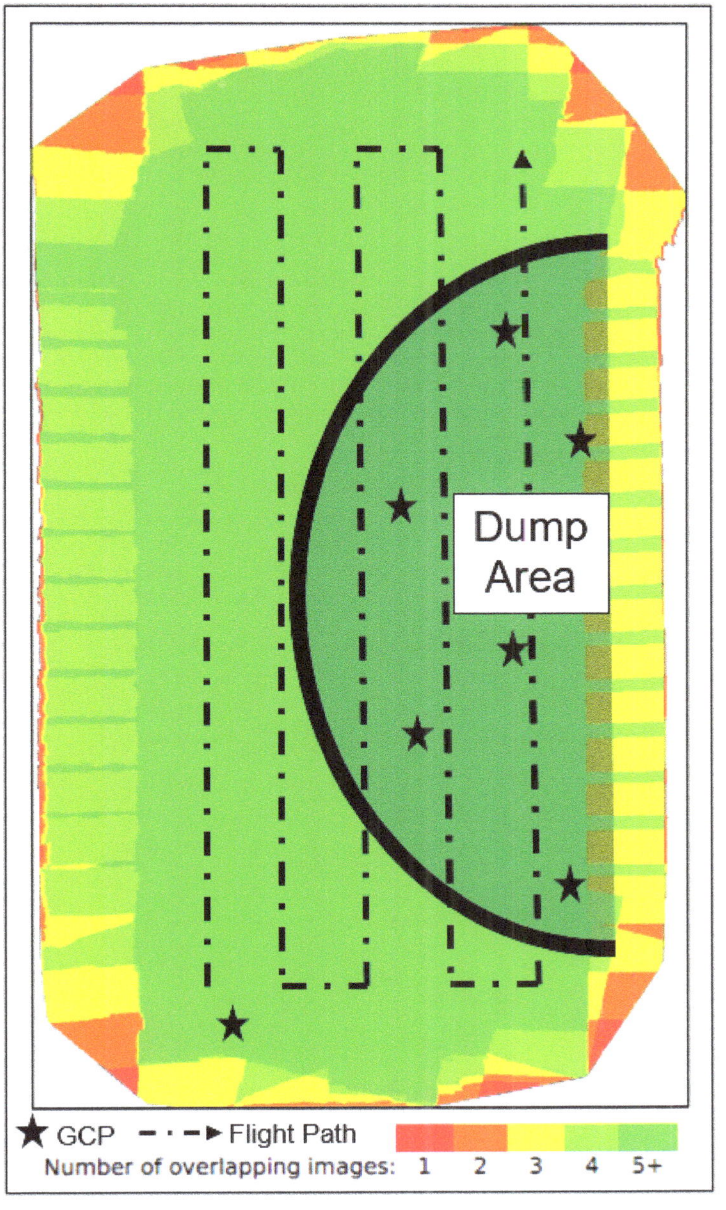

Figure 1. Top (XY Plane) view of a typical flight plan. Stars represent locations of ground control points (GCPs). The dashed line shows the flight path of the UAV. The shaded area is an approximation of the dump area. The color represents the number of overlapping images in accordance with the scale shown.

2.3. Point Cloud Analysis

Maptek™ PointStudio (2021.1 version, Maptek/KRJA Systems Inc, Golden, Colorado, USA) was used to model and analyze the resulting point cloud data. These data were imported as individual files and converted into triangulation solids. Then, they were analyzed for overlapping areas where dumps had occurred. Polygons were created around the dump areas, and the two triangulations were processed into extracted solids. These extracted solids are considered to be 3D volumetric HFMs of individual haul truck dump profiles within the simulation and modelling context described in the introduction. The reason these software tools, as well as the drone equipment, were used is because they were what was available and provided by the mine. The results section details the HFMs created by this study. All HFM files are available at the link, https://zenodo.org/record/5789951#.YgQiqJbMKUl.

2.4. Classification Method

Little taxonomic terminology yet exists for the information presented in this article. Classification involved qualitative analysis by the authors on the HFMs, with an emphasis on attributes hypothesized to be of interest to characterizing the final position and geometry of the dump profile. From the qualitative analysis, the categories and characteristics of the dumps were created, and each dump was assigned to the matching category. Data on the resulting classification are presented along with the statistical information of each category.

3. Results

3.1. UAV Flights and Field Work Results

Tables 2 and 3 show the root mean square (RMS) error values and the ground sampling distance (GSD) of the models computed from the drone flight data, respectively. Errors between the modelled coordinates of a GCP and its known survey coordinates are frequently used to represent the accuracy of a photogrammetry model [69].

Table 2. RMS Error Values by Coordinate Direction.

Coordinate Direction	Relative RMS Errors (m)				Absolute RMS Errors (m)			
	Minimum	Average	Maximum	Standard Deviation	Minimum	Average	Maximum	Standard Deviation
X	0.000565	0.011916	0.03008	0.008417	0.684995	2.410617	3.931733	0.790573
Y	0.000574	0.009329	0.032239	0.008137	1.128254	2.119136	3.641948	0.610887
Z	0.001044	0.017912	0.046895	0.011906	0.881072	7.083288	15.086508	4.262623

Table 3. GSD Values of the Photogrammetry Models.

GSD Values (cm/pixel)			
Minimum	Average	Maximum	Standard Deviation
1.72	2.2	2.52	0.26

RMS errors represent the quadratic means of these errors in the X, Y and Z coordinate directions. Both absolute and relative RMS error values are given. Absolute error represents the error in the coordinate location of the GCPs to their actual location on earth. Relative error represents the error in coordinate location of the GCPs as they relate to each other. Relative error is of more interest to this study, since the integrity of the HFMs is reliant on achieving a low relative error in the photogrammetry models.

Low ground sampling distance (GSD) values are required for accurate photogrammetry [69]. GSD is the physical distance represented between the centers of two adjacent pixels, and it can be estimated in advance of a flight based on the camera and flight details. With the drone flying at a 60 m height, a GSD of 1.64 cm/pixel is considered to be the lowest possible GSD value based on the camera specifications. Slight deviations due to perspective, the vibrations of the camera, blur, depth of field and other factors lead to variations in GSD at every point of a photogrammetric model. The GSD values calculated for the photogrammetry models of this study can be found in Table 3.

In Tables 2 and 3, the accuracy of the photogrammetry models is given. These accuracy values are important to consider, since they determine the granularity of the HFM models. The accuracy error of a given model will be at least the GSD value in each coordinate direction. Considering that the highest average GSD value of all flights was 2.52 cm/pixel, an accuracy error of at least 16 cm^3 (2.52 cm × 2.52 cm × 2.52 cm) to the volume of the HFM is expected to exist. Another way to confirm these accuracy errors is to multiply the RMS errors (Table 2) for each coordinate direction. The multiplication of the maximum relative errors for each coordinate direction, as shown in Table 2, gives 45.48 cm^3, which is roughly 2.8 times the accuracy error obtained by cubing the GSD. This discrepancy is consistent with other photogrammetry models that have been correctly scaled and reconstructed, which typically contain accuracy errors between one to three times that of the GSD value. In summary, the accuracy of the volumes generated from the photogrammetry models used in this study can generally be considered to be accurate to within 50 cm^3.

3.2. Point Cloud Analyses and Results

Of the 23 flights flown, 14 solids were extracted, from which 29 dump activities were identified and 28 were considered useable HFMs of the dumping process. Of these 28 HFMs, four types of dump profiles were determined to exist. The four dump types are named after their shape, as follows:

A. Oval,
B. Comet,
C. Rectangular,
D. Sloughed Heap.

Oval-type dump profiles are the most commonly occurring type of dump profile. These dump profiles are characterized by their oval shape when viewed from a vantage point normal to the dump face. These dumps have narrow ends at the crest and toe of the dump face, and a maximum width midway through the dump face. Comet dump profiles are characterized by a large volume near the base of the dump and a narrow trail of material extending upwards along the dump face. Rectangular dump profiles cover the entire dump face (or a large portion of it) evenly to a uniform width. Sloughed heap profiles occur when a portion of the material is not dumped over the edge of the berm, but rather on the floor of the upper level of the dump. This causes a portion of the material to bunch near the dump berm, and a part of it to slough over the edge of the dump.

Table 4 shows example grayscale renderings of what is observed in PointStudio for each of the four types of dumping profiles determined by a qualitative analysis of the dump HFMs. These example grayscale renderings are showcased via both front and side views. The front view shows the perspective facing the dump and in-line/parallel to the dump, with the dumping process proceeding from top to bottom. The side view represents the perspective perpendicular to the front view, with the dumping direction proceeding from top right to bottom left. Measurements of these profiles were performed using software functions found within PointStudio.

The HFM solids created in PointStudio were classified using the following variables: volume (in cubic meters), maximum height, length, width and thickness (in meters) and angle (in degrees) from bottom left to top right. Volume was determined by querying the properties of the HFM solids after their creation. Maximum height and maximum length were considered to be the vertical and horizontal legs of the right triangle formed by connecting the bottom left and top right points of the HFM solid. Maximum width was considered to be the distance between the two farthest horizontal points of the HFM solid from the front-view perspective. Thickness was considered to be the distance between the two farthest horizontal points of the HFM solid from the side-view perspective. Angle was considered to be the inner angle of the right triangle formed by connecting the bottom-left and top-right points of the HFM solid. These measurements are tabulated in Table 5. An additional visualization of the HFMs data containing all of the extracted solids is available in the data cache associated with this article. The link accessed to this data can be found in the "Supplementary Materials" section.

In total, 28 dumps were profiled and classified via qualitative analysis. Table 5 shows information from all of the dump profiles. With the exception of dumps 1, 2 and 3, all dumps were along the edge of a 30 m high dump crest. Table 5 reveals that large volume does not always indicate large maximum width, height or length. High angles do not indicate large volume, but seem to increase the maximum width, height or length. In the case of sloughed heaps, the low angle is due to the fact that it is not measuring to the top of the berm, but to the extent of the dump profile, which occurs on the floor of the upper level of the dump area.

Table 4. Example front and side views of grayscale renderings of the HFM solids created in PointStudio for each of the four dump profile types.

Dump Profile Type	Front View (Direction of Dumping ↓)	Side View (Direction of Dumping ←)
Oval 10 m	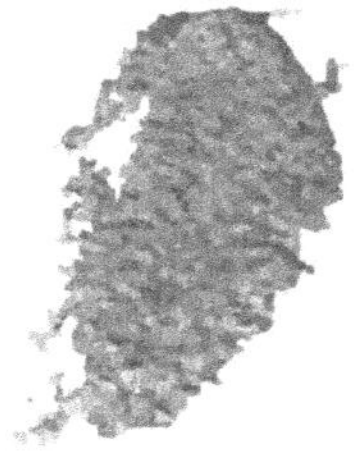	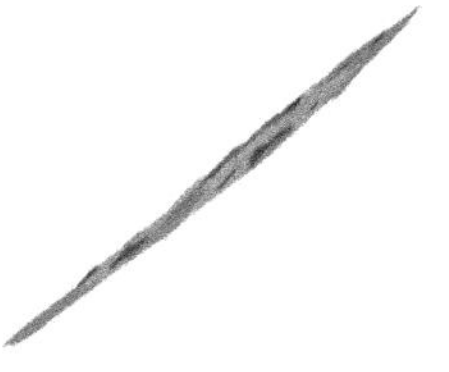
Comet 10 m		
Rectangular 10 m		
Sloughed Heap 10 m		

Table 5. Summary of Classification Data for All Dump Profiles.

Dump	Volume (m³)	Max. Width (m)	Max. Height (m)	Max. Length (m)	Angle (°)	Max. Thickness (m)	Shape
1 [2]	131	15	15	23	33	1.187	Oval
2 [2]	125 [1]	12	16	25	31	0.957	Rectangular
3 [2]	125 [1]	14	16	25	31	0.893	Rectangular
4	121 [1]	11	16	29	28	1.431	Sloughed Heap
5	121 [1]	14	30	44	36	0.368	Oval
6	121 [1]	13	31	43	36	1.207	Oval
7	155	20	20	32	33	1.016	Comet
8	119 [1]	22	26	39	34	1.123	Comet
9	119 [1]	17	18	26	35	1.256	Comet
10	119 [1]	11	32	46	34	1.006	Oval
11	134	20	28	40	35	0.887	Oval
12	125 [1]	15	21	35	31	1.675	Comet
13	125 [1]	19	19	29	33	2.011	Comet
14	137 [1]	16	19	27	29	1.053	Sloughed Heap
15	137 [1]	12	31	44	35	1.306	Oval
16	137 [1]	23	19	20	32	2.032	Comet
17	137 [1]	15	23	34	34	0.899	Oval
18	137 [1]	19	20	28	30	1.081	Sloughed Heap
19	128	20	21	30	36	0.953	Oval
20	138 [1]	16	28	39	31	1.000	Sloughed heap
21	138 [1]	16	25	36	35	1.202	Oval
22	117	11	7	16	12	2.062	Sloughed Heap
23	94	16	26	35	36	0.601	Oval
24	149 [1]	14	30	41	36	0.882	Rectangular
25	149 [1]	17	29	40	35	1.140	Oval
26	149 [1]	13	21	29	36	1.036	Rectangular
27	129 [1]	12	9	13	16	1.647	Sloughed Heap
28	129 [1]	17	26	37	35	1.033	Oval

[1] Volumes are averages from the total volume of the combined extracted solid containing the dump profiles. [2] These had a 15 m dump height.

In order to investigate the statistical differences among the classification data of each type, box plots of their data for each variable are shown in Figures 2–7. In the box plots, the black dot represents the mean average value within the data, the line represents the median value, the box edges represent 50% of the data between the first and third quartiles and the lines above and below the box represent the maximum and minimum values of the data.

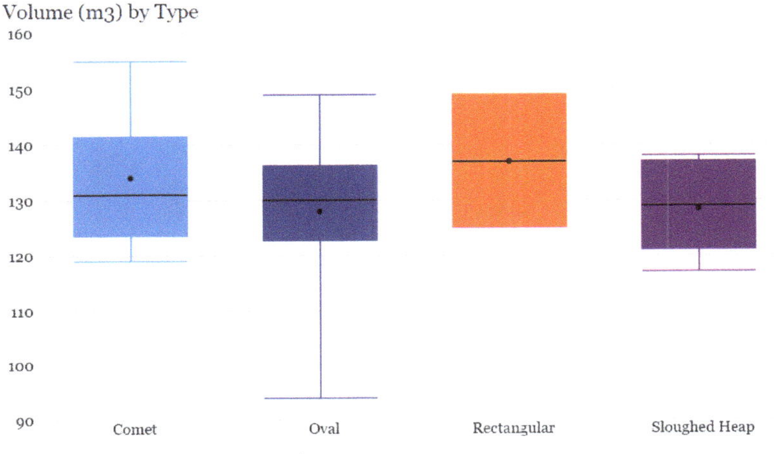

Figure 2. Box Charts of Volume (m³) by Dump Profile Shape.

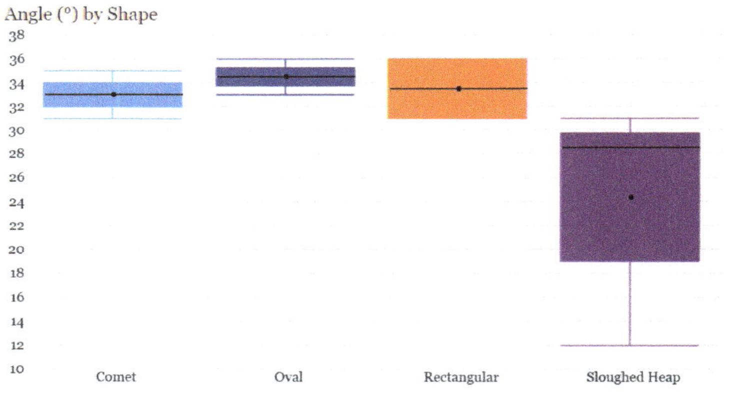

Figure 3. Box Charts of Angle (°) by Dump Profile Shape.

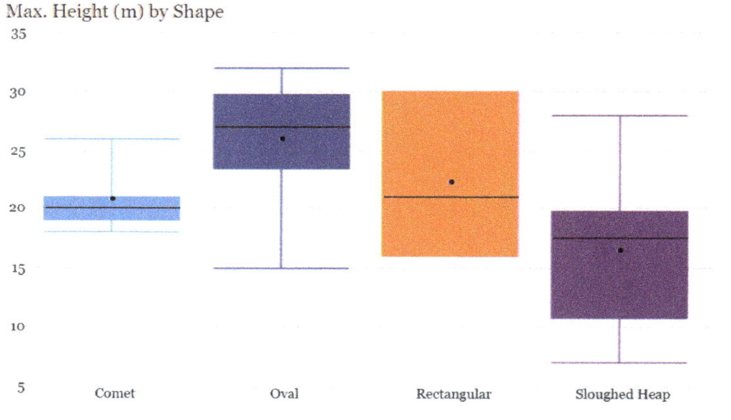

Figure 4. Box Charts of Max. Height (m) by Dump Profile Shape.

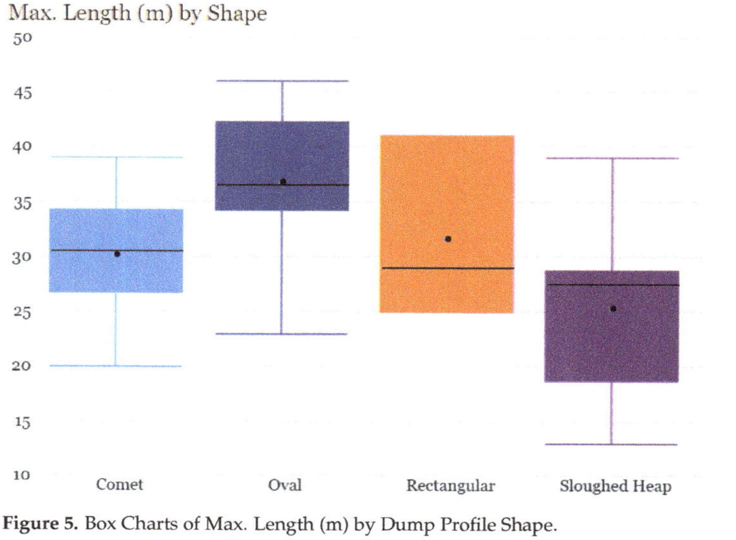

Figure 5. Box Charts of Max. Length (m) by Dump Profile Shape.

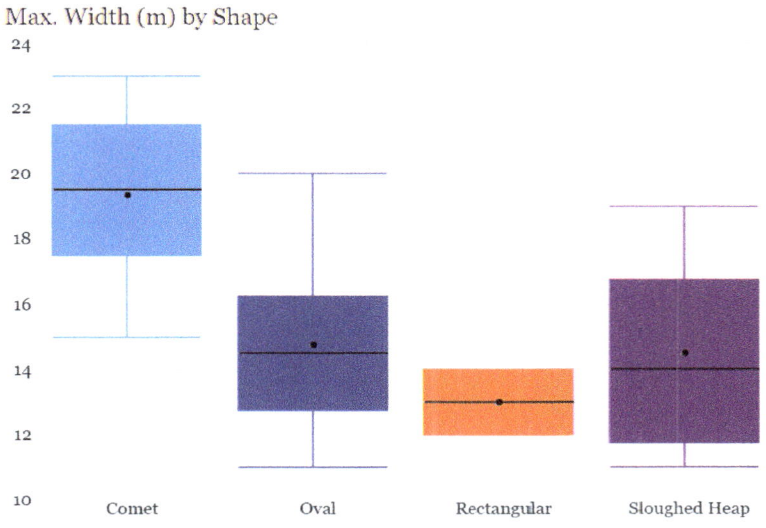

Figure 6. Box Charts of Max. Width (m) by Dump Profile Shape.

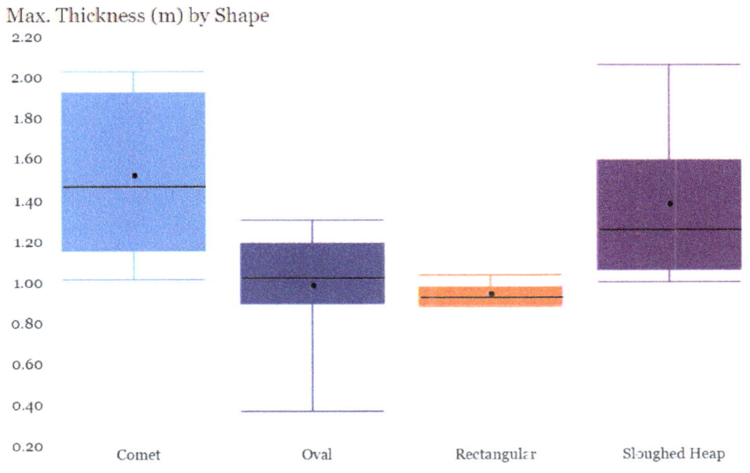

Figure 7. Box Charts of Max. Thickness (m) by Dump Profile Shape.

Figure 2 shows that the comet and rectangular dump types have a slightly higher average volume compared with the oval and sloughed heap types. Volume is influenced by how much material was loaded into the truck before dumping, and also by how much material in the dump face moved at the time of dumping. This increase in the average volume of material for the comet and rectangular dumps may be due to the fact that both dump types seem to involve the movement of additional material contained in the dump face.

Figure 3 shows a lot of similarity in the average angles of the comet, oval and rectangular dump types. Sloughed heap angles are lower because the material does not typically extend the full length of the dump face. Where the material of the sloughed heap extends along the dump face, the angle of the material matches the angles for the other dump types.

Figure 4 shows that the oval and rectangular dumps have the highest dump height. Dump height is a measurement of how much of the vertical dimension of the dump is covered by the dump profile. Sloughed heap and comet height values are lower because the material does not typically extend the full length of the dump face.

Figure 5 shows the same general differences between dump types as Figure 4. This is likely due to the fact that length is a measurement of how much of the horizontal dimension of the dump is covered by the dump profile. Length values are lower for sloughed heap- and comet-type dump profiles compared to the other dump types. This may be related to the same reason why height values for these types are also lower.

Figure 6 shows that comet dump profiles tend to be the widest. This may be the result of additional material from the dump face aggregating with the dump mass as it cascades, resulting in an increase in width. Rectangular dump profiles tend to have the lowest width values, which is interesting because they typically have the highest volume values, and this might be explained by a low amount of frictional resistance on the dump face compared to the cohesion of the dump mass.

As shown in Figure 7, the comet and sloughed heap dump profiles typically have a higher thickness than the oval and rectangular ones. Assuming that the dumps are of similar total volume, this is to be expected, since oval and rectangular dump profiles typically have higher height and length. Therefore, the dump material for oval and rectangular dumps is spread thinner across more surface area, which leads to less thickness. Comet dump profiles may also interact with loose material on the dump surface, and the resulting solid might include some of that material in the thickness, as it might with the total volume.

4. Discussion

4.1. Modelling Discussion

The HFMs presented in this paper offer a novel look at individual dumps. These HFMs can be used as a basis to calibrate future simulation models of dumps via the V-model for calibration described by Quist [63], Hofmann [66] and others. Figure 8 illustrates the V-model for simulation calibration and validation, which occurs over three levels. At the bottom level, validation is carried out by calibrating the individual parameters of particles in a laboratory setting. The second level involves calibration based on the aggregated behavior of the particles through multiple flow regimes. The third level compares simulated outcomes with real industrial-scale operations.

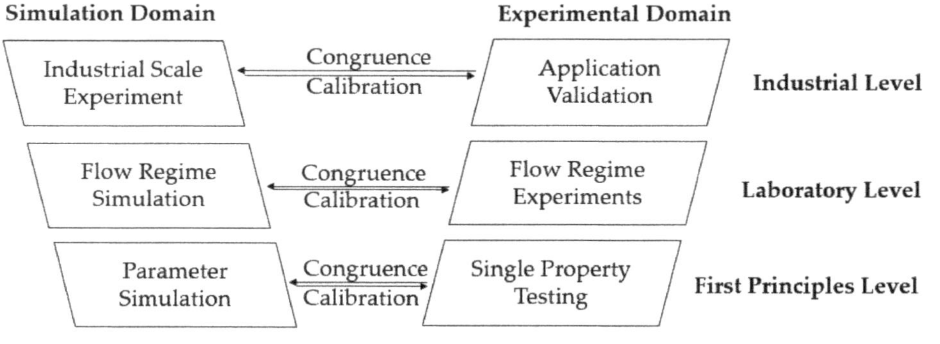

Figure 8. The V-model for DEM calibration and validation (adapted from [63]).

To the knowledge of the authors, no work has been conducted to validate dumping behavior at a real mine against simulated outcomes; however, much work has been performed to calibrate and validate ROM material simulations at lower levels of the V-model, particularly for mineral processing applications [70]. In instances such as dumping, where reality cannot be used to calibrate models, calibration involves adjusting parameters to fit HFMs instead. There is no mention of a method for creating HFMs of mine haul truck dumping activity in the literature. The method used for this study is easily achievable for any mine operation where UAVs and photogrammetry are presently in use.

4.2. Dump Profile Discussion

Even though this study found only four types of dump profiles, there may exist additional types. The factors that determine which type a dump will be categorized into are many. These factors likely fall into several categories, including the way the material is loaded into the truck, how the truck dumps the material, how the pre-existing dump face interacts with the dumped load and how the material behaves on its own. None of these factors were investigated in this paper

How the truck is loaded directly influences the volume of the resulting dump profile. It may also influence the determination of the dump profile type, forming either comet, oval or rectangular profiles. For example, if the truck is unevenly loaded with more material in the back than towards the front, this may cause the resulting dump profile to favor a comet format. However, the exact interplay between how the trucks were loaded and the resulting dump profiles remains unclear, and no information on truck loading was gathered during this study.

How the truck dumps the material clearly influences whether or not the dump profile becomes a sloughed heap or one of the other types. This is because if the truck dumps far from the crest of the dump face, it will create a sloughed heap. When the truck dumps against the crest of the dump face, the type of the resulting dump profile is either comet, oval or rectangular; however, it is unclear which one it will become from this information alone.

4.3. Terminology Discussion

The lack of terminology around dump profile behavior is one challenge to improving our understanding. The authors present new terminology here for the four types of dump shapes, but much more terminology is likely needed to fully describe and characterize dump profiles. This is made true by the fact that many dump profiles display characteristics of more than one type and, therefore, additional description may be required to fully classify them. For example, a sloughed heap may slough into a comet shape at the bottom of the dump, or an oval may have an extremely long rectangular section. There also needs to be reflection around how much effort should be spent creating terminology, since it is well known that the shape of mine rock piles is mainly based on topography [9].

4.4. Practicality Discussion

Admittedly, some findings in this study may not offer much practical applicability, especially considering the status quo of the mining industry. It is unknown whether the HFMs presented in this study represent the global characteristics of the dumping process across all mines, or whether they are limited to the mine used for the study. However, the method used for data capture was seamlessly incorporated into routine operation, and was practical from the standpoint of simplicity and ease of realization. Many mines are capable of measuring their own dump characteristics and creating HFMs for themselves. It is entirely probable that many other shapes of fallow land exist, and their discovery and classification can increase our understanding of the cascading process of ROM material.

Additionally, as has been previously stated, the findings of this study are practical for the purpose of verifying the accuracy of the simulation modelling of individual truck end dumps. Without an HFM to verify simulations against, there would be less confidence in the accuracy and relevance of the simulation. It is computationally intensive to simulate an entire dump area. Simulating individual dumps may allow for larger areas to be simulated with less computational power through the use of pseudo-particles [4].

Dumps generally conform to whatever shape the local topography provides, but knowledge of the common shapes provided via a routine dumping process for a particular ore at a given mine has historically been mostly speculatory. Speculation of this kind causes engineers and mine planners to place an unknown amount of dependance on operators that work with the material constantly to ensure the process is conforming to plan. In future autonomous mining scenarios, this dependance will not be allowed. Thus, digitally transforming the dumping process is hypothesized to support continuity in the transition towards autonomous mining. Without operators at the helm, there is little knowledge to ensure that dumping is occurring correctly. Additionally, without a basic reference, there would be no way to determine the performance quality of the autonomous equipment. The fact that autonomous equipment will be covered in sensors to map the work area continuously will mean very little, unless it is known what conformity should actually looks like.

As another hypothesis, there might be a relationship of practical synergy between predictive modelling and UAV surveys. UAV surveys have become increasingly ubiquitous at mines, and can accomplish tasks to a level of quality unachievable by traditional survey crews within a greatly reduced timeframe. They could be made to occur at such frequent intervals as to overwhelm mining engineers and long-range planners. The HFMs described in this article demonstrate the parameterization of the dumping process, which could facilitate the training and validation of predictive models to help automate design conformity to UAV surveys of rock piles, thereby decreasing the cognitive load placed on the domain experts.

5. Conclusions

HFMs of dump profiles for 28 dumps were created. These HFMs show characteristic behaviors classifiable into four types, named comet, oval, rectangular and sloughed heap. These classifications may make it easier to examine the dumping process as a whole. While more terminology and modelling will be required to gain a complete understanding of the dumping process, the HFMs examined here provide a basis to articulate new terminology and calibrate new modelling.

Since further investigation is required, some recommendations for future study in this area include:

- Investigate factors that determine the classification of a given dump profile;
- Isolate additional variables that influence the cascading behavior of ROM from haul trucks;
- Simulate and calibrate particle modelling using HFMs;
- Validate and test the ability to accurately simulate and predict the dump characteristics beforehand;
- Correct for the difference between GPS coordinates recorded as dump locations and the true centroid coordinates of the dumped material;
- Develop constraints and map FMS data to rock piles;
- Prove the accuracy of these mapping/modelling techniques through a robust sampling campaign;
- Adapt this or a similar method for dozers and other equipment that frequently handle material at dumps and stockpiles;
- Review and analyze correlated phenomena (bulk phenomena in heap leaches, stratification, slope stability, etc.);
- Incorporate simulated predictions into a larger mine-to-mill optimization model.

Supplementary Materials: The dump geometries can be downloaded at: https://zenodo.org/record/5789951#.YgQiqJbMKUl.

Author Contributions: Conceptualization, A.Y.; Data curation, A.Y.; Formal analysis, A.Y.; Funding acquisition, W.P.R.; Investigation, A.Y.; Methodology, A.Y.; Project administration, W.P.R.; Resources, W.P.R.; Software, A.Y.; Supervision, W.P.R. All authors have read and agreed to the published version of the manuscript.

Funding: This research received no external funding.

Data Availability Statement: Data supporting the reported results can be found via the link in the Supplementary Materials.

Conflicts of Interest: The authors declare no conflict of interest.

References

1. Erkayaoglu, M.; Dessureault, S. Improving mine-to-mill by data warehousing and data mining. *Int. J. Min. Reclam. Environ.* **2019**, *33*, 409–424. [CrossRef]
2. Adel, G.; Kojovic, T.; Thornton, D. *Mine-to-Mill Optimization of Aggregate Production*; Virginia Polytechnic Inst. and State Univ.(Virginia Tech), Blacksburg: Virginia, VA, USA, 2006.
3. Valery, W.; Duffy, K.; Jankovic, A. Mine to Mill Optimization. In *SME Mineral Processing and Extractive Metallurgy Handbook*; Kawatara, K., Young, C.A., Eds.; Society for Mining, Metallurgy, and Exploration (SME): Englewood, CO, USA, 2019; pp. 335–343.
4. Servin, M.; Vesterlund, F.; Wallin, E. Digital Twins with Distributed Particle Simulation for Mine-to-Mill Material Tracking. *Minerals* **2021**, *11*, 524. [CrossRef]
5. Holmberg, K.; Kivikytö-Reponen, P.; Härkisaari, P.; Valtonen, K.; Erdemir, A. Global energy consumption due to friction and wear in the mining industry. *Tribol. Int.* **2017**, *115*, 116–139. [CrossRef]
6. Hartman, H.L.; Mutmansky, J.M. *Introductory Mining Engineering*; John Wiley & Sons: Hoboken, NJ, USA, 2002.
7. Young, A.; Rogers, W.P. Modelling Large Heaped Fill Stockpiles Using FMS Data. *Minerals* **2021**, *11*, 636. [CrossRef]
8. McLemore, V.T.; Fakhimi, A.; van Zyl, D.; Ayakwah, G.F.; Anim, K.; Boakye, K.; Ennin, F.; Felli, P.; Fredlund, D.; Gutierrez, L.A.F.; et al. *Literature Review of Other Rock Piles: Characterization, Weathering, and Stability*; Questa Rock Pile Weathering Stability Project; New Mexico Bureau of Geology and Mineral Resources: Socorro, NM, USA, 2009.
9. Zahl, E.G.; Biggs, F.; Boldt, C.M.K.; Connolly, R.E.; Gertsch, L.; Lambeth, R.H.; Stewart, B.M.; Vickery, J.D. Waste Disposal and Contaminant Control. In *SME Mining Engineering Handbook*; Darling, P., Ed.; Society for Mining, Metallurgy and Exploration Inc.: Littleton, CO, USA, 1992; pp. 1170–1180.
10. Siddiqui, F.; Shah, S.; Behan, M. Measurement of Size Distribution of Blasted Rock Using Digital Image Processing. *J. King Abdulaziz Univ. Eng. Sci.* **2009**, *20*, 81–93. [CrossRef]
11. Badroddin, M.; Bakhtavar, E.; Khoshrou, H.; Rezaei, B. Efficiency of standardized image processing in the fragmentation prediction in the case of Sungun open-pit mine. *Arab. J. Geosci.* **2013**, *6*, 3319–3329. [CrossRef]
12. Maerz, N.H. Online Fragmentation Analysis: Achievements in the Mining Industry. In Proceedings of the 7th Annual ICAR Symposium, Austin, TX, USA, 19–21 April 1999.
13. Bao, Y.; Han, X.; Chen, J.; Zhang, W.; Zhan, J.; Sun, X.; Chen, M. Numerical assessment of failure potential of a large mine waste dump in Panzhihua City, China. *Eng. Geol.* **2019**, *253*, 171–183. [CrossRef]
14. Poulsen, B.; Khanal, M.; Rao, A.M.; Adhikary, D.; Balusu, R. Mine overburden dump failure: A case study. *Geotech. Geol. Eng.* **2014**, *32*, 297–309. [CrossRef]
15. Mhaske, S.; Kapoor, I.; Pathak, K.; Kayet, N. Slope Stability Analysis of the Overburden Dump of Meghahatuburu Iron Ore Mines in Singhbhum Region of India. In Proceedings of the International Field Exploration and Development Conference 2019, Xi'an, China, 16–18 October 2019; Springer: Singapore, 2020.
16. McCoy, J.T.; Auret, L. Machine learning applications in minerals processing: A review. *Miner. Eng.* **2019**, *132*, 95–109. [CrossRef]
17. Bartos, P.J. Is mining a high-tech industry? *Investigations into innovation and productivity advance. Resour. Policy* **2007**, *32*, 149–158. [CrossRef]
18. Sánchez, F.; Hartlieb, P. Innovation in the mining industry: Technological trends and a case study of the challenges of disruptive innovation. *Min. Metall. Explor.* **2020**, *37*, 1385–1399. [CrossRef]
19. Hustrulid, W.A.; Kuchta, M.; Martin, R.K. *Open Pit Mine Planning and Design, Two Volume Set & CD-ROM Pack, [CD-ROM]*; CRC Press: Leiden, The Netherlands, 2013.
20. Casey, J. AngloGold Ashanti Provides Obuasi Wnderground Mining Update. Basic Materials & Resources Monitor Worldwide (Amman, Jordan). 2021. Available online: https://www.globalminingreview.com/mining/30092021/anglogold-ashanti-provides-obuasi-underground-mining-update/ (accessed on 28 January 2022).
21. Kamoa-Kakula Copper Project Mines and Stockpiles 269,000 Tonnes of Ore from the Kakula and Kansoko Mines in December, at an Average Grade of 5.36% Copper, Including 55,000 Tonnes Grading 8.75% Copper from the High-Grade Centre of the Kakula Mine, in ENP Newswire. 2021. Available online: https://link.gale.com/apps/doc/A647556674/ITOF?u=marriottlibrary&sid=bookmark-ITOF&xid=71f68e94/ (accessed on 26 January 2022).
22. Yannopoulos, J.C. Leaching Low-Grade Gold Ores. In *The Extractive Metallurgy of Gold*; Springer: Boston, MA, USA, 1991; pp. 115–136.
23. Yi, R. Optimum determination of sub-grade stockpiles in open pit mines. *AUSIMM* **1988**, *4*, 57–62.
24. Investigators at Colorado School of Mines Detail Findings in Operations Research (Practical Performance of an Open Pit Mine Scheduling Model Considering Blending and Stockpiling). *Computer Weekly News*. 2020, p. 285. Available online: https://link.gale.com/apps/doc/A619469258/ITOF?u=marriottlibrary&sid=bookmark-ITOF&xid=73be3380 (accessed on 26 January 2022).
25. Arnoldi, M. AfriTin Suspends Mining in Namibia, Continues Processing. 2020. Available online: https://www.engineeringnews.co.za/article/afritin-suspends-mining-in-namibia-continues-processing-2020-03-31 (accessed on 26 January 2022).
26. Dutta, S. Record 132 Million Tonne Coal Stockpiles Cast Shadow over New Mines. 2021. Available online: https://timesofindia.indiatimes.com/business/india-business/record-132-million-tonne-coal-stockpiles-cast-shadow-over-new-mines/articleshow/81973823.cms (accessed on 26 January 2022).

27. Arnoldi, M. ARMs Iron-Ore Mines Allowed to Continue Loading from Stockpile. 2020. Available online: https://www.miningweekly.com/article/arms-iron-ore-mines-allowed-to-continue-loading-from-stockpile-2020-04-03 (accessed on 26 January 2022).
28. Lundin Gold Inc. Lundin Gold Restarts Operations at Fruta Del Norte and Provides 2020 Outlook. 2020. Available online: https://www.newswire.ca/news-releases/lundin-gold-restarts-operations-at-fruta-del-norte-and-provides-2020-outlook-832174232.html (accessed on 26 January 2022).
29. Legal Monitor Worldwide. Endeavour Silver Resumes Operations at Three Mexico Mines. 2020. Available online: https://link.gale.com/apps/doc/A625237491/ITOF?u=marriottlibrary&sid=bookmark-ITOF&xid=703ae665 (accessed on 26 January 2022).
30. Legal Monitor Worldwide. Lundin Mining Announces Resumption of Full Production Rates at Chapada Mine. 2020. Available online: https://link.gale.com/apps/doc/A646443468/ITOF?u=marriottlibrary&sid=bookmark-ITOF&xid=d2753530 (accessed on 26 January 2022).
31. Basic Materials & Resources Monitor Worldwide (Amman, Jordan). Endeavour Expects Mining Activities to Resume at Boungou in The Coming Weeks. 2020. Available online: https://link.gale.com/apps/doc/A634219247/ITOF?u=marriottlibrary&sid=bookmark-ITOF&xid=2681c7ac (accessed on 26 January 2022).
32. Basic Materials & Resources Monitor Worldwide (Amman, Jordan). Trevali Provides Update on Extension of Government Declaration of National Emergency in Peru. 2020. Available online: https://link.gale.com/apps/doc/A618915153/ITOF?u=marriottlibrary&sid=bookmark-ITOF&xid=05ffadcf (accessed on 26 January 2022).
33. Plus Company Updates. Gold Resource Corporation's Don David Gold Mine Begins Process of Ramping up Activities, Update's 2021 Guidance. 2021. Available online: https://link.gale.com/apps/doc/A674749128/ITOF?u=marriottlibrary&sid=bookmark-ITOF&xid=2e950161 (accessed on 26 January 2022).
34. Legal Monitor Worldwide. Metro Mining Delays Bauxite Hills Expansion Amid Market Uncertainties. 2020. Available online: https://link.gale.com/apps/doc/A624952426/ITOF?u=marriottlibrary&sid=bookmark-ITOF&xid=52e3f3cc (accessed on 26 January 2022).
35. Carter, R.A. Staying on Top of Stockpile Management. *Eng. Min. J.* **2018**, *219*, 58.
36. Legal Monitor Worldwide. Clever Solutions, Right Equipment Saves Re-mining Project. 2021. Available online: https://link.gale.com/apps/doc/A668313798/ITOF?u=marriottlibrary&sid=bookmark-ITOF&xid=9dae0354 (accessed on 26 January 2022).
37. M & A Navigator. Deal Snapshot: Trident Resources Acquires Copper Royalty for USD 5m. 2020. Available online: https://link.gale.com/apps/doc/A627983091/ITOF?u=marriottlibrary&sid=bookmark-ITOF&xid=019a144b (accessed on 26 January 2022).
38. Legal Monitor Worldwide. Prospect Mining Studio Selects Startups to Pilot Sustainability Solutions. 2021. Available online: https://link.gale.com/apps/doc/A671464583/ITOF?u=marriottlibrary&sid=bookmark-ITOF&xid=c5bd619a (accessed on 26 January 2022).
39. Braun, R.L.; Lewis, A.E.; Wadsworth, M.E. In-place leaching of primary sulfide ores: Laboratory leaching data and kinetics model. *Metall. Trans.* **1974**, *5*, 1717–1726. [CrossRef]
40. Bartlett, R.W. Simulation of ore heap leaching using deterministic models. *Hydrometallurgy* **1992**, *29*, 231–260. [CrossRef]
41. Davis, G.B.; Ritchie, A.I.M. A model of oxidation in pyritic mine wastes: Part 1 equations and approximate solution. *Appl. Math. Model.* **1986**, *10*, 314–322. [CrossRef]
42. Ghorbani, Y.; Becker, M.; Mainza, A.; Franzidis, J.-P.; Petersen, J. Large particle effects in chemical/biochemical heap leach processes–A review. *Miner. Eng.* **2011**, *24*, 1172–1184. [CrossRef]
43. Wang, L.; Yin, S.; Deng, B. Understanding the Effect of Stepwise Irrigation on Liquid Holdup and Hysteresis Behavior of Unsaturated Ore Heap. *Minerals* **2021**, *11*, 1180. [CrossRef]
44. Lizama, H.M. How copper dump leaching works. *Miner. Eng.* **2021**, *171*, 107075. [CrossRef]
45. Van Staden, P.J.; Petersen, J. The effects of simulated stacking phenomena on the percolation leaching of crushed ore, Part 2: Stratification. *Miner. Eng.* **2019**, *131*, 216–229. [CrossRef]
46. Benito, J.G.; Ippolito, I.; Vidales, A.M. Novel aspects on the segregation in quasi 2D piles. *Powder Technol.* **2013**, *234*, 123–131. [CrossRef]
47. Zhang, S.; Liu, W. Application of aerial image analysis for assessing particle size segregation in dump leaching. *Hydrometallurgy* **2017**, *171*, 99–105. [CrossRef]
48. Plus Company Updates. Cobalt Blue Holdings Limited: COB Executes MOU with State of Queensland: Recovery of Cobalt from Mine Waste. 2021. Available online: https://link.gale.com/apps/doc/A687700113/ITOF?u=marriottlibrary&sid=bookmark-ITOF&xid=99e7db65 (accessed on 26 January 2022).
49. Basic Materials & Resources Monitor Worldwide (Amman, Jordan). Prairie River Minerals and NRRI Pilot Scram Mining Operation. 2021. Available online: https://link.gale.com/apps/doc/A674990833/ITOF?u=marriottlibrary&sid=bookmark-ITOF&xid=25ea5faa (accessed on 26 January 2022).
50. Mena Report. Australia: Territory Iron Shipment Leaves Darwin Port. 2021. Available online: https://link.gale.com/apps/doc/A666537309/ITOF?u=marriottlibrary&sid=bookmark-ITOF&xid=f12334b3 (accessed on 26 January 2022).
51. Plus Company Updates. Red River Resources Limited: Red River Announces Maiden Gold Resource for Hillgrove Stockpile. 2020. Available online: https://link.gale.com/apps/doc/A617069866/ITOF?u=marriottlibrary&sid=bookmark-ITOF&xid=ee3db896 (accessed on 26 January 2022).

52. ENP Newswire. Equus Mining Limited-High Grade Pegaso Drill Results Confirm Potential of Mineralisation along Trend from Historic Mines. 2021. Available online: https://link.gale.com/apps/doc/A662116991/ITOF?u=marriottlibrary&sid=bookmark-ITOF&xid=0c36356d (accessed on 26 January 2022).
53. Plus Company Updates. Elim Mining Announces $140 Million Cumulative Free Cash Flow from the Cactus Mine Stockpile Preliminary Economic Assessment. 2020. Available online: https://link.gale.com/apps/doc/A617122264/ITOF?u=marriottlibrary&sid=bookmark-ITOF&xid=bcc4ef16 (accessed on 26 January 2022).
54. ENP Newswire. Providence Gold Drilling Stockpile. 2020. Available online: https://link.gale.com/apps/doc/A642311201/ITOF?u=marriottlibrary&sid=bookmark-ITOF&xid=9dd90cde (accessed on 26 January 2022).
55. ENP Newswire. Southern Empire Provides Oro Cruz Project Historical Mine Dumps Assay Results and Preliminary Cyanidation Metallurgical Test Results. 2021. Available online: https://link.gale.com/apps/doc/A666663259/ITOF?u=marriottlibrary&sid=bookmark-ITOF&xid=c7fe179b (accessed on 26 January 2022).
56. Legal Monitor Worldwide. Eclipse Metals Targets Mining Licence to Breathe New Life into Historic Ivittuut Cryolite Mine. 2021. Available online: https://link.gale.com/apps/doc/A666122579/ITOF?u=marriottlibrary&sid=bookmark-ITOF&xid=70d6d24a (accessed on 26 January 2022).
57. Young, A.; Rogers, P. A review of digital transformation in mining. *Min. Metall. Explor.* **2019**, *36*, 683–699. [CrossRef]
58. Vial, G. Understanding digital transformation: A review and a research agenda. *J. Strateg. Inf. Syst.* **2019**, *28*, 118–144. [CrossRef]
59. Hinings, B.; Gegenhuber, T.; Greenwood, R. Digital innovation and transformation: An institutional perspective. *Inf. Organ.* **2018**, *28*, 52–61. [CrossRef]
60. Job, A.; McAree, P.R. Three Case Studies on the Implementation of New Technology in the Mining Industry. In Proceedings of the Iron Ore Conference, Perth, Australia, 24–26 July 2017.
61. Chaulya, S.; Prasad, G.M. *Sensing and Monitoring Technologies for Mines and Hazardous Areas: Monitoring and Prediction Technologies*; Elsevier: Amsterdam, The Netherlands, 2016.
62. Francioni, M.; Salvini, R.; Stead, D.; Giovannini, R.; Riccucci, S.; Vanneschi, C.; Gullì, D. An integrated remote sensing-GIS approach for the analysis of an open pit in the Carrara marble district, Italy: Slope stability assessment through kinematic and numerical methods. *Comput. Geotech.* **2015**, *67*, 46–63. [CrossRef]
63. Johannes Quist, M.E. Framework for DEM Model Calibration and Validation. In Proceedings of the 14th European Symposium on Comminution and Classification, Gothenburg, Sweden, 7–10 September 2015.
64. Durst, P.J.; Goodin, C. High fidelity modelling and simulation of inertial sensors commonly used by autonomous mobile robots. *World J. Model. Simul.* **2012**, *8*, 172–184.
65. George, A.D.; VanLoon, R.A. High-Fidelity Modelling and Simulation of Myrinet System Area Networks. *Int. J. Model. Simul.* **2001**, *21*, 40–50. [CrossRef]
66. Hofmann, M. On the Complexity of Parameter Calibration in Simulation Models. *J. Def. Model. Simul.* **2005**, *2*, 217–226. [CrossRef]
67. Vermeer, W.; Hjorth, A.; Jenness, S.M.; Brown, C.H.; Wilensky, U. Leveraging Modularity During Replication of High-Fidelity Models: Lessons from Replicating an Agent-Based Model for HIV Prevention. *J. Artif. Soc. Soc. Simul.* **2020**, *23*, 7. [CrossRef]
68. Zhao, S. 3D Real-Time Stockpile Mapping and Modelling with Accurate Quality Calculation Using Voxels. Ph.D. Thesis, University of Adelaide, Adelaide, Australia, 2016.
69. Siqueira, H.L.; Marcato, J.; Matsubara, E.; Eltner, A.; Colares, R.A.; Santos, F.M. The Impact of Ground Control Point Quantity on Area and Volume Measurements with UAV SFM Photogrammetry Applied in Open Pit Mines. In Proceedings of the 2019 IEEE International Geoscience and Remote Sensing Symposium (IGARSS), Yokohama, Japan, 28 July–2 August 2019.
70. Tavares, L.M. A review of advanced ball mill modelling. *KONA Powder Part. J.* **2017**, *34*, 106–124. [CrossRef]

mining

MDPI

Article

Rock Fragmentation Prediction Using an Artificial Neural Network and Support Vector Regression Hybrid Approach

Richard Amoako [1,*], Ankit Jha [1] and Shuo Zhong [2]

1 Department of Mining Engineering and Management, South Dakota School of Mines and Technology, Rapid City, SD 57701, USA; ankitkrjha@gmail.com
2 Strayos Inc., Buffalo, NY 14203, USA; shuo@strayos.com
* Correspondence: ramoako.kwesi@gmail.com

Abstract: While empirical rock fragmentation models are easy to parameterize for blast design, they are usually prone to errors, resulting in less accurate fragment size prediction. Among other shortfalls, these models may be unable to accurately account for the nonlinear relationship that exists between fragmentation input and output parameters. Machine learning (ML) algorithms are potentially able to better account for the nonlinear relationship. To this end, we assess the potential of the multilayered artificial neural network (ANN) and support vector regression (SVR) ML techniques in rock fragmentation prediction. Using geometric, explosives, and rock parameters, we build ANN and SVR models to predict mean rock fragment size. Both models yield satisfactory results and show higher performance when compared with the conventional Kuznetsov model. We further demonstrate an automated means of analyzing a varied number of hidden layers for an ANN using Bayesian optimization in the Keras Python library.

Keywords: rock fragmentation prediction; machine learning; Kuz–Ram model; fragmentation models; fragment size distribution; artificial neural network; support vector regression; blasting; open-pit mines

Citation: Amoako, R.; Jha, A.; Zhong, S. Rock Fragmentation Prediction Using an Artificial Neural Network and Support Vector Regression Hybrid Approach. *Mining* **2022**, *2*, 233–247. https://doi.org/10.3390/mining2020013

Academic Editors: Juan M Menéndez-Aguado, Oscar Jaime Restrepo Baena and Jessica M. Smith

Received: 2 March 2022
Accepted: 20 April 2022
Published: 24 April 2022

1. Introduction

Rock fragmentation is the process by which rock is broken down into smaller size distributions by mechanical tools or by blasting. The resulting fragment size distribution may be characterized by a histogram showing the percentage of sizes of particles, or as a cumulative size distribution curve [1]. The primary means of rock fragmentation in mining is blasting. A good blast produces a size distribution that is well suited to the mining system it feeds, maximizes saleable fractions, and enhances the value of saleable material [2]. Blasting efficiently saves significant amounts of money that would otherwise be spent on secondary blasting [3]. It also yields significant savings on the costs of downstream comminution processes, i.e., crushing and grinding.

The results of a blast depend on several parameters, which are broadly categorized as controllable and uncontrollable [4,5]. Controllable parameters can be varied by the blasting engineer to adjust the outcome of blasting operations. Controllable parameters can be grouped into geometric, explosives, and time parameters. Geometric parameters include drill hole diameter, hole depth, charge length, spacing, burden, and stemming height. Explosives parameters include the type of explosive, explosive strength and energy, powder factor, and priming systems. Time parameters include delay timing and initiation sequence. A blasting engineer's ability to change these controllable parameters dynamically in response to as-drilled information is critical to achieving good fragmentation [3]. The uncontrollable parameters constitute the geological and geotechnical properties of the rock mass. These parameters are inherent, and thus, cannot be varied to adjust blasting outcomes. They include rock strength, rock-specific gravity, joint spacing and condition, presence and depth of water, and compressional stress wave velocity [6]. Though these parameters cannot be varied by the blasting engineer, adequately accounting for them in a blast design helps to achieve good fragmentation. Figure 1 is a bench blast profile showing a variety of design parameters.

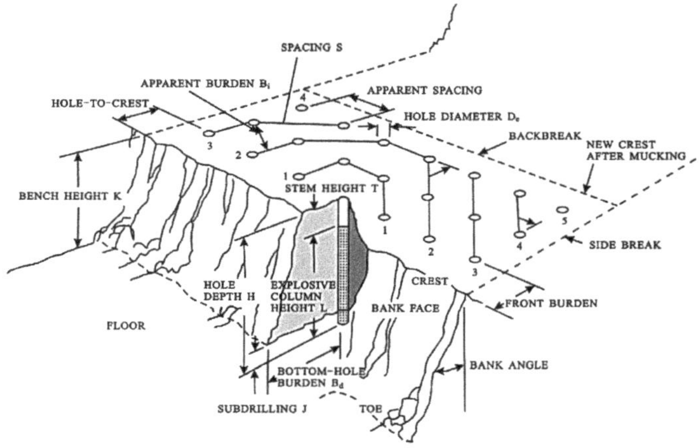

Figure 1. Blast design terminology [5].

Several studies have sought to predict fragment size distribution based on the parameters used in blast design. The accurate prediction will give blasting engineers control over the outcome of blasting operations. Consequently, engineers will know which controllable parameters to modify, and to what extent the modification should be. Having an accurate prediction model leads to good post-blast results, and this comes with enhanced loader and excavator productivity along with numerous downstream benefits. However, the prediction exercise proves to be challenging considering that numerous parameters influence fragmentation. Additionally, the rock mass may be heterogeneous and/or anisotropic in its structures of weakness. To this end, it is impossible to develop a predictive tool solely based on theoretical and mechanistic reasoning [5]. Researchers have thus mostly resorted to empirical techniques in predicting the outcome of fragmentation, with the Kuz–Ram being the most widely used. The empirical models are favored and widely used in daily blasting operations because they are easily parameterized. A major shortfall, however, with the empirical methods is that certain significant parameters are not accounted for, and this leads to less accurate results. Cunningham [2], notes that essential parameters omitted by empirical techniques include rock properties and structure, e.g., joint spacing and condition, detonation behavior, and mode of decking. Other parameters include blast dimensions and edge effects from the borders of the blast. Over the years, researchers have modified existing models and formulated new ones in an attempt to improve prediction accuracy. While this has contributed to significant improvement, none of the ensuing models incorporate all the important parameters, and accuracy is still of concern. In some instances, highly simplified or inappropriate procedures were used for estimating the properties of structural weakness in the rock mass [5]. Furthermore, the relationship between fragmentation input and output parameters is highly nonlinear, and empirical models may not be well suited for such modeling.

To this end, researchers, in recent years, have sought to implement machine learning (ML) techniques for fragmentation prediction. The objective was to capture as much of the inherent nonlinearity using limited input parameters and subsequently improve accuracy. Kulatilake et al. [5] and Shi et al. [7] have respectively exploited the potential of using artificial neural network (ANN) and support vector regression (SVR) for this purpose, and have achieved satisfactory results. ANN and SVR are machine learning techniques that are proven to possess high nonlinearity-recognition properties. However, ANN models in the rock fragmentation literature were limited to only one hidden layer, and do not exploit the potential of the multilayered network (ANN with more than one hidden layer), which could potentially lead to achieving higher accuracy. In this research, we implement SVR and a variety of multilayered ANN for predicting mean fragment size.

Machine learning (ML) is a branch of artificial intelligence (AI) that allows computer systems to improve their performance at a task through experience (learning) for the purpose of predicting future outcomes [7,8]. It is a multidisciplinary field that relies significantly on specialized subject areas such as probability and statistics, and control theory. ML techniques are broadly classified as supervised and unsupervised learning. Supervised learning is concerned with predicting an outcome given a set of input data. It does so by making use of the already established relationship between representative sets of input and output data that were used for model training. Unsupervised learning is concerned with data segmentation based on pattern recognition. Unsupervised ML techniques can infer patterns from data without reference to known outcomes. They are useful for discovering the underlying structure of a given data set. The rock fragmentation problem is a regression problem that is suited to tools of supervised machine learning such as multivariate regression analysis, artificial neural network (ANN), and support vector regression (SVR). The last two comprise algorithms that are more robust to nonlinear relationships between input and output data [5,9]. They are thus considered in this study since rock fragmentation input and output parameters are nonlinearly related.

2. Preliminary Background

We provide a fundamental explanation of the machine learning techniques used in this study. The section describes the architecture of the artificial neural network and support vector regression.

2.1. *Artificial Neural Network (ANN)*

Artificial neural network (ANN) is a machine learning technique that is inspired by the way the biological neural system works, such as how the brain processes information [7,8,10]. Information processing in ANN involves many highly interconnected processing elements known as neurons that work together to solve specific problems. The learning process involves adjustments to the synaptic connections existing between the neurons [7,11]. In the biological neural system, a neuron consists of a cell body, known as soma, an axon, and dendrites. The axon sends signals, and the dendrites receive these signals. A synapse connects an axon to a dendrite. Depending on the signal it receives, a synapse might increase or decrease electrical potential. An ANN consists of a number of neurons similar to human biological neurons. These neurons are known as units and are connected by weighted links that transmit signals from one neuron to the other [7,12]. The output signal is transmitted through the neuron's outgoing connection, which is analogous to the axon in the biological neuron. The outgoing connection splits into a number of branches that transmit the same signal. The outgoing branches terminate at the incoming connections (analogous to dendrites) of other neurons in the network [7].

An ANN has three types of neurons, and these are known as input, hidden, and output neurons. They are stacked in layers, and receive input from preceding neurons or external sources, and use this to compute an output signal using an activation function. The activation function is a mathematical formula for determining the output of a neuron based on the neuron's weighted inputs. The output signal is then propagated to succeeding neurons. While this is ongoing, the ANN adjusts its weights in order to record an acceptable minimal error between input variables and the final output variable(s) [13]. The complexity of the ANN architecture makes it well suited for solving both linear and nonlinear problems [10]. Advancement in computational power has enhanced its use in the fields of engineering, industrial process control, medicine, risk management, marketing, finance, communication, and transportation.

2.2. *Suport Vector Regression (SVR)*

Support vector regression (SVR) is a type of supervised machine learning that is based on statistical learning theory [14]. Just like the ANN, SVR is efficient at modeling nonlinearly related variables and does well at solving both classification and regression problems. It works by nonlinearly mapping, i.e., transforming, a given data set into a higher dimensional feature space, and then solving a linear regression problem in this feature space [9,15]. That is, it seeks to predict a single output variable ($\hat{y}$) as a function of n input variables (x) using a function $f(x)$ that has at most ε deviation from the actual values (y) for all the training data [16]. Equation (1) expresses this function in its simplest form as a linear relationship [9]:

$$f(x) = b + w \cdot \varphi(x) \tag{1}$$

In Equation (1), the function $\varphi(x)$ denotes the high dimensional kernel-induced feature space. Kernel refers to the mathematical function used in the data transformation process. Different kernels are available for use in SVR analysis. They include the linear, polynomial, radial basis function (rbf), and sigmoid kernels. Parameter w in Equation (1) is a weight vector, and b is a bias term. Both w and b are calculated by minimizing a regularized cost function. Figure 2 is a graphical representation of the SVR concept. The $\pm\varepsilon$ deviation from the actual values (y) can be described as a tube that contains the sample data with a certain limit ε [16]. This implies that the function $f(x)$ is constrained by the $\pm\varepsilon$ limits to form a tube that represents the data set with the expected deviations.

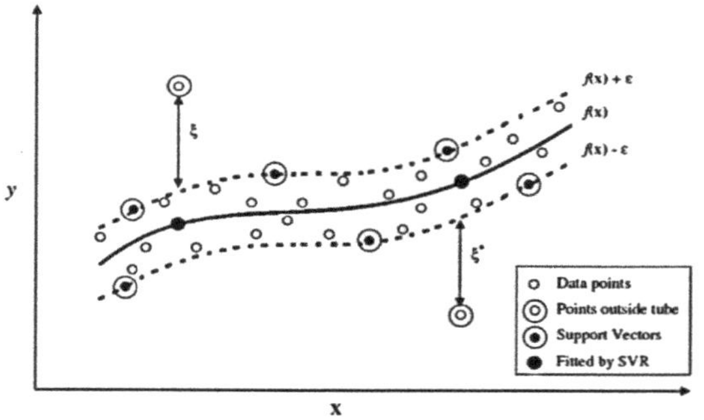

Figure 2. Graphical representation of support vector regression [17].

3. Literature Review

The ability to accurately predict fragment size distribution from a given blast design will give blasting engineers control over the outcome of blasting operations. Engineers will be able to identify which controllable parameters to modify, and to what extent the modification should be. To this end, several studies have sought to predict fragment size distribution based on the parameters used in blast design. These studies have resulted in empirical prediction models, with the Kuz–Ram being the commonest model in use. Others include the CZM, two-component model (TCM), Kuznetsov–Cunningham–Ouchterlony (KCO), SveDeFo, and Larson's equation [4,18]. The reliance on empirical models stems from the complexity that comes with the attempt to develop explicit theoretical and mechanistic equations to predict the outcome of fragmentation [2,4,5]. This complexity is primarily attributed to the fact that there are so many parameters that affect a blast, coupled with geological heterogeneity [5,9].

The Kuz–Ram model is essentially a three-part model consisting of a modified version of the Kuznetsov equation, the Rossin–Rammler equation, and the Cunningham uniformity index. The parameters defined by these equations constitute the output of the prediction model [4]. The Kuznetsov equation is for predicting mean fragment size (X_{50}), and the original version is given by Kuznetsov [19] as:

$$X_{50} = A\left(\frac{V}{Q}\right)^{0.8} Q^{0.167} \tag{2}$$

In Equation (2), X_{50} is the mean fragment size (cm); A is the rock factor (7 for medium hard rocks, 10 for hard but highly fissured rocks, 13 for very hard, weakly fissured rocks); V is the rock volume (m^3); and Q is the weight of TNT (kg) equivalent in energy to the explosive charge in one borehole. A shortfall of the equation is that the rock mass categories it defines are very wide, and thus need more precision [5]. Cunningham [20,21] provides a modified version of the equation as follows:

$$X_{50} = AK^{-0.8}Q^{\frac{1}{6}}\left(\frac{115}{RWS}\right)^{\frac{19}{20}} \tag{3}$$

In Equation (3), A is the rock factor, and varies between 0.8 and 22 depending on hardness and structure; K is the powder factor, defined as the weight of explosive, in kg, per cubic meter of rock; Q is the mass, in kg, of the explosive in the hole; and RWS is the weight strength relative to ANFO (115 is the RWS of TNT).

The role of the Rosin Rammler equation is to estimate the complete fragmentation distribution. For a given mesh size or screen opening, X, this equation is able to estimate the percentage of fragments retained. It is given as [22]:

$$R_x = \exp^{-}\left(\frac{X}{X_c}\right)^n \tag{4}$$

where R_x is the proportion of fragments larger than the mesh size X (cm), and X_c is the characteristic fragment size (cm). The characteristic size is one through which 63.2% of the materials pass. If the characteristic size and the uniformity index are known, a size distribution curve can be plotted for the rock fragments [18]. The curve is plotted as percentage passing vs. mesh size. The former is obtained by subtracting R_x from one. Equation (4) can be rewritten to make direct use of the mean fragment size, X_{50}, as follows [20,21]:

$$R_x = \exp^{-0.693}\left(\frac{X}{X_{50}}\right)^n \tag{5}$$

From Equations (4) and (5), the characteristic size can be deduced as:

$$X_c = \frac{X_{50}}{0.693^{\frac{1}{n}}} \tag{6}$$

The third part of the Kuz–Ram model is the uniformity index, developed by Cunningham through several investigations which involved consideration of the effects of blast geometry, hole diameter, burden, spacing, hole length, and drilling accuracy [4]. This equation is given as [20,21]:

$$n = \left(2.2 - \frac{14B}{d}\right)\sqrt{\left(\frac{1+\frac{S}{B}}{2}\right)}\left(1 - \frac{W}{B}\right)\left(abs\left(\frac{BCL - CCL}{L}\right) + 0.1\right)^{0.1}\frac{L}{H} \tag{7}$$

where B is the burden (m); S is the spacing (m); d is the hole diameter (mm); W is the standard deviation of drilling precision (m); L is the charge length (m); BCL is the bottom charge length (m); CCL is the column charge length (m); and H is the bench height (m). Equation (7) is multiplied by 1.1 when using a staggered pattern. The value of n is essential in determining the shape of the size distribution curve, and is usually between 0.7 and 2. High values indicate uniform sizing, while low values indicate a wide range of sizes, including both oversize and fines [18,23]. Equations (3), (5), and (7) are what constitute the typical Kuz–Ram model.

Cunningham [2] makes modifications in the model twenty years on, mainly as a result of the introduction of electronic delay detonators. This leads to what is now known in the literature as the modified Kuz–Ram model. The adjustments by Cunningham incorporate the effects of inter-hole delay and timing scatter. The changes also incorporate correction factors for the rock factor and uniformity index. These changes lead to the modification of Equations (3) and (7) as follows [2]:

$$X_{50} = AA_T K^{-0.8} Q^{\frac{1}{6}} \left(\frac{115}{RWS}\right)^{\frac{19}{20}} C(A) \tag{8}$$

$$n = n_S \sqrt{\left(2 - \frac{30B}{d}\right)} \sqrt{\left(\frac{1 + \frac{S}{B}}{2}\right)} \left(1 - \frac{W}{B}\right) \left(\frac{L}{H}\right)^{0.3} C(n) \tag{9}$$

where A_T is a timing factor for the effect of inter-hole delay, $C(A)$ is a correction factor for the rock factor, n_s is the uniformity factor for the effect of timing scatter, and $C(n)$ is a correction factor for the uniformity index. Thus, the modified Kuz–Ram model comprises Equations (5), (8) and (9).

A major shortfall of the Kuz–Ram model is the underestimation of fines. Extensions to the model have, thus, emerged with the objective of improving the prediction of fines. The CZM and TCM are such models [18]. Kanchibotla, Valery, and Morrell [24] address the issue of fines via the CZM model, which provides fragment distribution based on the coarse and fine parts of the muck pile. The authors note that during blasting, two different mechanisms control rock fragmentation, i.e., tensile fracturing and compressive-shear fracturing. Tensile fracturing produces coarse fragments, while compressive fracturing produces the fines. The model predicts the coarser part of the size distribution using the Kuz–Ram model. The size distribution of the finer part is predicted by modifying the values of n and X_c in the Rosin–Rammler equation. Djordjevic [25] develops a two-component model (TCM) based on the same mechanisms of failure captured by Kanchibotla et al. [24] in their work. The model utilizes experimentally determined parameters from small-scale blasting, and parameters of the Kuz–Ram model to obtain an improved prediction of fragment size distribution.

Ouchterlony [26] develops the KCO model which ties in the Kuz–Ram, CZM, and TCM models. The KCO model replaces the original Rosin–Rammler equation with the Swebrec function to predict rock fragment size distribution. The replacement stems from the author's recognition that the Rosin–Rammler curve has limited ability to follow the various distributions from blasting. The Swebrec function proves to be more adaptable and is able to predict fines better. The model is given by Equations (10) and (11) as follows [26]:

$$P(x) = \frac{1}{[1 + f(x)]} \tag{10}$$

$$f(x) = \left[\frac{\ln\left(\frac{X_{max}}{X}\right)}{\ln\left(\frac{X_{max}}{X_{50}}\right)}\right]^b \tag{11}$$

where $P(x)$ is the percentage of fragments passing a given mesh size, X; X_{max} is the upper limit of fragment size; X_{50} is the mean fragment size; and b is the curve undulation parameter. Just like the Rosin–Rammler model, the Swebrec function has the mean fragment size (X_{50}) as its central parameter but introduces an upper limit to fragment size (X_{max}). While the aforementioned extensions to the Kuz–Ram model improve the distribution of fines, they introduce yet another factor into a predictive model that is already somewhat extended [2].

With the advancement in computational power, attention is being drawn to the use of machine learning (ML) in rock fragmentation prediction. Over the last decade, researchers have used multivariate regression (MVR) analysis, artificial neural network (ANN), and support vector regression (SVR) to predict fragment size distribution. In their work, Hudaverdi, Kulatilake, and Kuzu [27] use MVR analysis to develop prediction equations for the estimation of the mean particle size of muck piles. They develop two different equations based on rock stiffness. The equations incorporate blast design parameters (i.e., burden, spacing, bench height, stemming, and hole diameter) expressed as ratios, explosives parameters (i.e., powder factor), and rock mass properties (i.e., elastic modulus and in situ block size). Comparative analysis involving results of the prediction equations, Kuznetsov empirical equation, and the actual values prove the capability of the proposed models in offering satisfactory results. The authors make use of a diverse database (the largest ever used in research at the time) representing blasts conducted in different parts of the world. This makes their prediction models robust to a wide range of blast design parameters and rock conditions.

Building upon the work of Hudaverdi et al. [27], Kulatilake et al. [5] developed MVR and ANN models for the same set of data used in the former authors' work. The authors train a single hidden layer neural network model to predict the mean particle size for each of two groups of data, as distinguished by the rock stiffness. The authors perform extensive analysis to determine the optimum number of neurons for the hidden layer. Comparative analysis reveals that the MVR and ANN models perform better than the conventional Kuznetsov model. Shi et al. [9] build upon the work of Kulatilake et al. [5] by exploiting the potential of using support vector regression (SVR) for predicting rock fragmentation. Using the same data set as the previous authors, Shi et al. [9] develop an SVR model for predicting mean fragment size. They compare the results of the SVR model with those of ANN, MVR, Kuznetsov, and the actual values. The comparison shows that SVR is capable of providing acceptable prediction accuracy.

The effectiveness of prediction models is assessed via comparative analysis involving post-blast measurement. Post-blast measurement techniques have been developed over the years for determining the true fragment size after a blast was completed. An accurate predictive model will record insignificant deviation from the true fragment size. The available techniques for measuring fragmentation output can be classified as direct and indirect [3]. The direct methods include sieve analysis, boulder count, and direct measuring of fragments. The most accurate method of determining fragmentation is to sieve the whole muck pile. However, because muck piles are large, the use of sieving and the other direct methods can be tedious, time-consuming, and costly [5]. Thus, they are not practicable for muck pile fragment distribution. They can, however, be used for smaller amounts of fragment materials, and for very special purposes [3].

The indirect methods of fragment size measurement include digital image processing, and measurement of parameters, which can be correlated to the degree of fragmentation [3]. Digital image processing involves the use of sophisticated software and hardware for measuring fragment size. It is the latest fragmentation analysis tool and has largely replaced the conventional methods. The use of this tool comprises the following steps: image capturing of muck pile, image scaling, image filtering, image segmentation, binary image manipulation, measurement, and stereometric interpretation [5]. Though quick and cost-effective, this tool has some challenges. Non-uniform lighting, shadows, and a large range of fragment sizes can make fragment delineation very difficult. Another challenge is the overestimation of fines since the computer treats all undigitized voids between the fragments as fines. Thus, to obtain accurate estimation, a correction must be applied. Additionally, the wide variations in size may require different scales of calibration [5,28].

4. Data and Methodology

This section discusses the data and methods employed in this study. The data set comprises 102 blasts. Using this data set, we develop a multilayered artificial neural network and support vector regression models that satisfactorily predict mean rock fragment size.

4.1. Data Source and Description

The data set used in this work is obtained from the blast database compiled by Hudaverdi et al. [27], and subsequently used by Kulatilake et al. [5] and Shi et al. [9]. The compilation consists of blast data from various mines around the world. The data, therefore, represents a diverse range of blast design parameters and rock formations. Having such a diverse range of data is good for the purpose of this study, i.e., training machine learning models for prediction. The implication here is that the predictive ability of the ensuing models would span a wide variety of rock formations. The compilation by Hudaverdi et al. [27] represents one of the largest and most diverse blast data collections in the literature, and thus fits the purpose of this study.

Table 1 shows a sample of the data. A summary of the individual research projects from which Hudaverdi et al. [27] compiled the data is provided hereafter. Blasts with labels "Rc", "En", and "Ru" are from research by Hamdi, Du Mouza, and Fleurisson [29], and Aler, Du Mouza, and Arnould [30] at the Enusa and Reocin mines in Spain. The Enusa Mine is an open-pit uranium mine in a schistose with moderate to heavily folded formation. The Reocin Mine is an open pit and underground zinc mine. Blasts designated "Mg" are from a study by Hudaverdi [31] at the Murgul Copper Mine, an open-pit mine in northeastern Turkey. Those designated "Mr" are from a study by Ouchterlony et al. [28] at the Mrica Quarry in Indonesia. The rock formation is mainly andesite. Blasts with the "Sm" label are from an open-pit coal mine in Soma Basin, in Western Turkey [32]. Blasts labeled "Db" are from the Dongri-Buzurg open-pit manganese mine in Central India. The rock formation is generally micaceous schist and muscovite schist [33]. Blasts labeled "Ad" and "Oz" are, respectively, from the Akdaglar and Ozmert quarries of the Cendere basin in northern Istanbul. Rock formation at both quarries is sandstone [27].

Table 1. Sample blast data [5,9,27–33].

ID	S/B	H/B	B/D	T/B	Pf ($\frac{kg}{m^3}$)	X_b (m)	E (Gpa)	X_{50} (m)
En1	1.24	1.33	27.27	0.78	0.48	0.58	60	0.37
En2	1.24	1.33	27.27	0.78	0.48	0.58	60	0.37
En3	1.24	1.33	27.27	0.78	0.48	1.08	60	0.33
Rc1	1.17	1.5	26.2	1.08	0.33	0.68	45	0.46
Rc2	1.17	1.5	26.2	1.12	0.3	0.68	45	0.48
Rc3	1.17	1.58	26.2	1.22	0.28	0.68	45	0.48
Mg1	1	2.67	27.27	0.89	0.75	0.83	50	0.23
Mg2	1	2.67	27.27	0.89	0.75	0.78	50	0.25
Mg3	1	2.4	30.3	0.8	0.61	1.02	50	0.27
Ru1	1.13	5	39.47	1.93	0.31	2	45	0.64
Ru2	1.2	6	32.89	3.67	0.3	2	45	0.54
Ru3	1.2	6	32.89	3.7	0.3	2	45	0.51
Mr1	1.2	6	32.89	0.8	0.49	1.67	32	0.17
Mr2	1.2	6	32.89	0.8	0.51	1.67	32	0.17
Mr3	1.2	6	32.89	0.8	0.49	1.67	32	0.13
Db1	1.25	3.5	20	1.75	0.73	1	9.57	0.44
Db2	1.25	5.1	20	1.75	0.7	1	9.57	0.76
Db3	1.38	3	20	1.75	0.62	1	9.57	0.35
Sm1	1.25	2.5	28.57	0.83	0.42	0.5	13.25	0.15
Sm2	1.25	2.5	28.57	0.83	0.42	0.5	13.25	0.19
Sm3	1.25	2.5	28.57	0.83	0.42	0.5	13.25	0.23
Ad1	1.2	4.4	28.09	1.2	0.58	0.77	16.9	0.15

Table 1. *Cont.*

ID	S/B	H/B	B/D	T/B	Pf ($\frac{kg}{m^3}$)	X_b (m)	E (Gpa)	X_{50} (m)
Ad2	1.2	4.8	28.09	1.2	0.66	0.56	16.9	0.17
Ad3	1.2	4.8	28.09	1.2	0.72	0.29	16.9	0.14
Oz1	1	2.83	33.71	1	0.48	0.45	15	0.27
Oz2	1.2	2.4	28.09	1	0.53	0.86	15	0.14
Oz3	1.2	2.4	28.09	1	0.53	0.44	15	0.14

The data set features blast design parameters that can be categorized as geometric, explosives, and rock parameters. The geometric parameters include burden, B (m), spacing, S (m), stemming, T (m), hole depth, H (m), and hole diameter, D (m). These are represented in the data set as ratios and include hole depth to burden (H/B), spacing to burden (S/B), burden to hole diameter (B/D), and stemming to burden (T/B) ratios. The powder factor, Pf ($\frac{kg}{m^3}$), represents the explosives parameter and shows the distribution of explosives in the rock. The elastic modulus, E (GPa), and the in situ block size, X_b (m), represent the rock parameters. Specifically, in situ block size represents the rock mass structure, while the elastic modulus represents the intact rock properties [27]. In effect, a total of seven rock fragment size prediction parameters are in the data set, and these will constitute the input parameters (independent variables) for the SVR and ANN models. The data set also features a post-blast parameter, i.e., X_{50}(m), which is the actual mean fragment size. This will be the output parameter (dependent variable) to be predicted by the models. Table 2 shows the summary statistics of the seven input parameters and the mean fragment size for the entire data set.

Table 2. Summary statistics.

	Variable	Minimum	Maximum	Mean	Standard Deviation
Input	S/B	1	1.75	1.20	0.11
	H/B	1.33	6.82	3.46	1.60
	B/D	17.98	39.47	27.23	4.91
	T/B	0.5	4.67	1.27	0.69
	Pf (kg/m^3)	0.22	1.26	0.55	0.24
	X_b (m)	0.29	2.35	1.16	0.48
	E (Gpa)	9.57	60	30.18	17.52
Output	X_{50} (m)	0.12	0.96	0.31	0.18

4.2. Model Development

Support vector regression (SVR) and artificial neural network (ANN) models are built for a total of 102 blasts. We split the data into training and test sets comprising 90 and 12 blasts, respectively. The test set has Kuznetsov predictions matching the actual fragment size. This is for the purpose of comparative assessment of results. The data set is scaled within the range 0–1 since the parameters have different orders of magnitude. The scaling is performed using the MinMaxScaler function of the Scikit-learn Python library [34]. The SVR and ANN models are built using the Scikit-learn and Keras Python libraries, respectively [34,35].

4.2.1. SVR Modeling

Using Scikit-learn, we develop and train a support vector regression model for prediction. The modeling process involves iterating over several combinations of the following support vector hyper-parameters: regularization (C), epsilon (ε), and kernel (k). Four kernels are considered for modeling, i.e., radial basis function (rbf), polynomial (poly), sigmoid, and linear. Twenty-five different values of C are considered in the interval [1:10], and twenty-seven different values of ε are considered in the interval [1×10^{-6}:0.3]. This yields a total of 2700 combinations of hyper-parameters, each representing a unique SVR model. The process of searching for the optimal combination of these hyper-parameters (adjustable parameters which control the support vector) is known as hyper-parameter tuning. To aid with this process, the GridSearchCV function in Scikit-learn is used [34]. It involves building SVR models using each of these hyper-parameter combinations and subsequently using cross-validation to assess model performance. We adopt the five-fold cross-validation technique. This means that for each hyper-parameter combination, the data are split into five folds. The hyper-parameter combination undergoes five runs of model training, and during each run, a distinct fold (one-fifth of the training data) is set aside for validation purposes. The final score assigned to the hyper-parameter combination is the average validation score from the five runs. This process is repeated for all other hyper-parameter combinations. We retrieve the best performing combination of hyper-parameters, and these are C = 5.25, ε = 0.04, and kernel = rbf. The final SVR model is thus built using these hyper-parameters.

In this study, retrieval of the best performing combination is based on the mean squared error (MSE) scoring metric. The MSE is a statistical metric that provides a means of assessing performance between two or more models. For each model, the MSE measures the average squared difference between the actual and predicted values. A perfect model would yield an MSE of zero, signifying that the actual values are perfectly predicted by the model, i.e., there is no error in prediction. In machine learning, the best-performing model among alternatives will be the one with MSE closest to zero. We show the MSE values for selected hyper-parameter combinations for the training and test data in Figure 3. From the figure, we observe that models with rbf kernels have better generalization abilities in respect of unseen, real-world data, i.e., data not included in the training process. This is represented by the test data. The best-performing model retrieved from the hyper-parameter tuning is of the rbf kernel type. It yields the lowest MSE value for the test data.

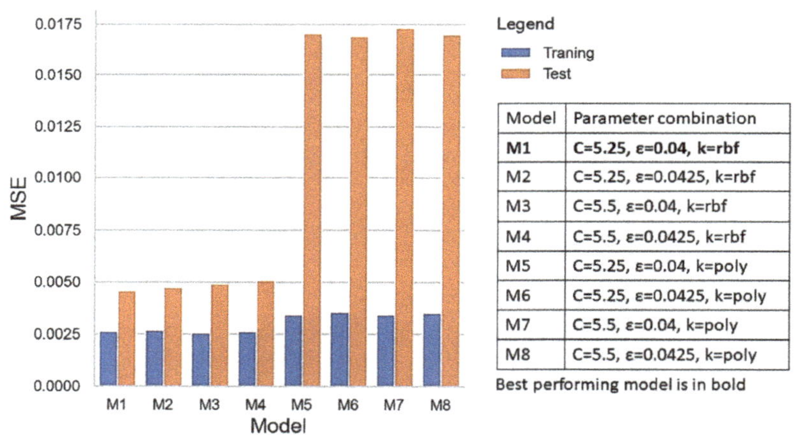

Figure 3. MSE plot for selected SVR hyper-parameter combinations.

4.2.2. ANN Modeling

Using Keras, we develop a variety of multilayered ANNs with up to four hidden layers for prediction. In each instance, hyper-parameter tuning is performed to obtain an optimal number of neurons (units) for the hidden layers under consideration. In all cases, the input and output layers have fixed neurons, being seven and one, respectively. These represent the seven input parameters, and the output parameter (X_{50}), which we seek to predict. Figure 4 is a schematic representing the general architecture of the ANNs used in this study.

For each instance of hidden layers, hyper-parameter tuning is performed using the Bayesian optimization object in Keras [35]. The process involves iterating over several combinations of neurons for a given instance of hidden layers and returning the combination that yields the best performance. This process can be very cumbersome and time-consuming when carried out manually. The use of Bayesian optimization saves time by automating the search process for the best combination of neurons for a given number of hidden layers. During the search process, 20% of the training data is set aside for validation purposes using the MSE scoring metric. The remaining data are used for training, and this involves running 1500 epochs to yield an acceptable reduction in prediction error.

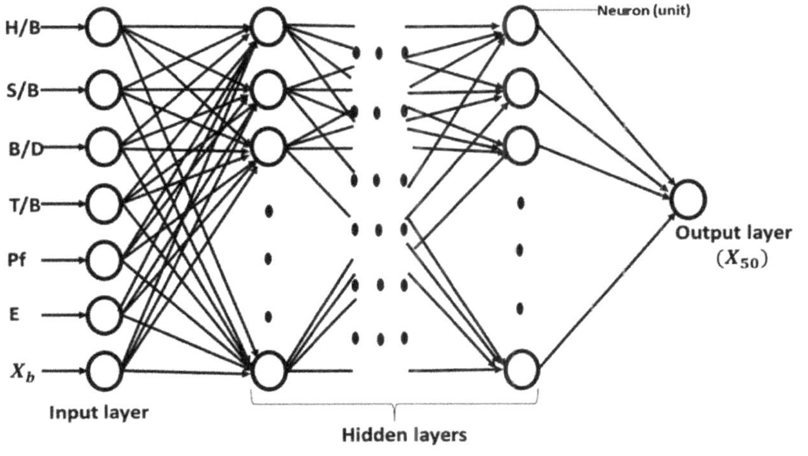

Figure 4. ANN architecture for rock fragmentation prediction.

Table 3 shows the results for the various hidden layers considered. For each instance of hidden layers, the table shows the optimal number of neurons returned via hyper-parameter tuning. The neural network with four hidden layers is selected as the final ANN model. This is based on the test scores, which represent the ability of the models to generalize to unseen, real-world data. The four-hidden-layer architecture has the lowest test score.

Table 3. Optimal neurons for hidden layers.

Number of Hidden Layers	Optimal Neurons for Hidden Layers	MSE for Test Data	Selected Model
1	90	0.0059	
2	25-BN-45	0.0039	
3	60-195-190	0.0040	
4	115-40-180-35	0.0031	✓

In the second configuration of hidden layers, the batch normalization (BN) technique serves to control model overfitting, so as to improve model generalization in respect of unseen, real-world data. Batch normalization applies a transformation that maintains the mean output close to zero and the output standard deviation close to 1, thereby standardizing the inputs to a given layer [35]. We show the performance of selected hyper-parameter combinations for the various hidden layer instances in Figure 5. The figure shows how the final ANN model (M8) compares with other models from the hyper-parameter tuning exercise. Model M5 has the worst generalization performance while model M8 has the best generalization performance.

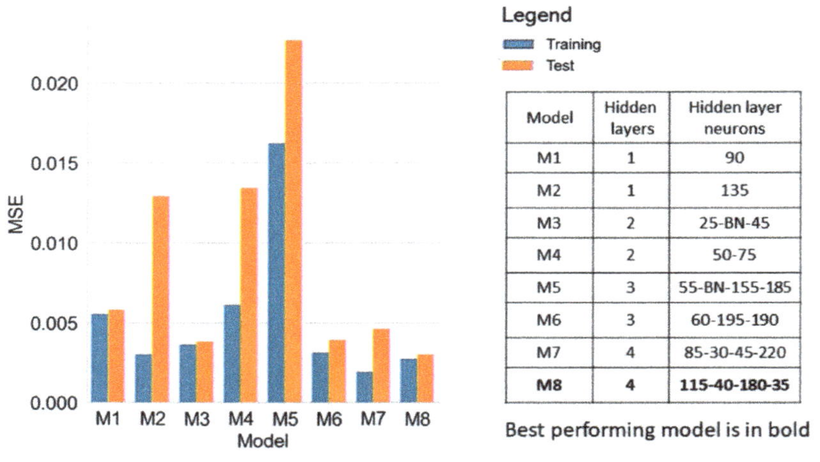

Figure 5. MSE plot for selected ANN hyper-parameter combinations.

5. Results and Discussion

Through hyper-parameter tuning, we obtain the final SVR and ANN models. For the purpose of assessing model generalization, we subject these models to testing. The test data set comprises 12 blasts; these are not used for training. The performance of the model on this data shows how well it will perform when deployed in the real world. Table 4 shows the performance of the final models on the training and test sets using the mean squared error (MSE) as a scoring metric.

Table 4. Model performance.

Model	Mean Squared Error (MSE)	
	Training	**Test**
SVR (C = 5.25, ε = 0.04, kernel = rbf)	0.0026	0.0044
ANN (115-40-180-35)	0.0028	0.0031

For the purpose of comparative assessment, the Kuznetsov empirical technique, i.e., Equation (3), is used to predict the mean rock fragment size for the test data. Test results obtained for the ANN and SVR models are compared with those for the Kuznetsov technique and the actual values. Table 5 and Figure 6 show the results for all three modeling techniques. It is observed that the ANN model records the least error while the Kuznetsov records the highest error. The coefficient of determination (r^2) measures the proportion of the variation in the dependent variable (mean fragment size) that is accounted for by its relationship with the independent variables. It ranges between zero and one. A model with r^2 closer to one is said to be reliable in predicting the dependent variable. The foregoing indicates that the ANN and SVR models are better able to model the relationship between the dependent and independent variables than the Kuznetsov empirical model. They show superior performance to the Kuznetsov as a result of their inherent ability to model complex, nonlinear relationships, such as exist between rock fragment size and blast design parameters.

Table 5. Results for test data.

Blast Number	Mean Fragment Size (m)			
	Actual	Predictions		
		ANN	SVR	Kuznetsov
1	0.47	0.44	0.38	0.48
2	0.64	0.68	0.64	0.71
3	0.44	0.38	0.41	0.42
4	0.25	0.25	0.25	0.33
5	0.20	0.15	0.14	0.27
6	0.35	0.21	0.52	0.09
7	0.18	0.19	0.19	0.38
8	0.23	0.17	0.18	0.22
9	0.17	0.17	0.19	0.25
10	0.21	0.21	0.20	0.12
11	0.20	0.21	0.19	0.13
12	0.17	0.24	0.26	0.23
Coefficient of determination (r^2)		0.87	0.81	0.58

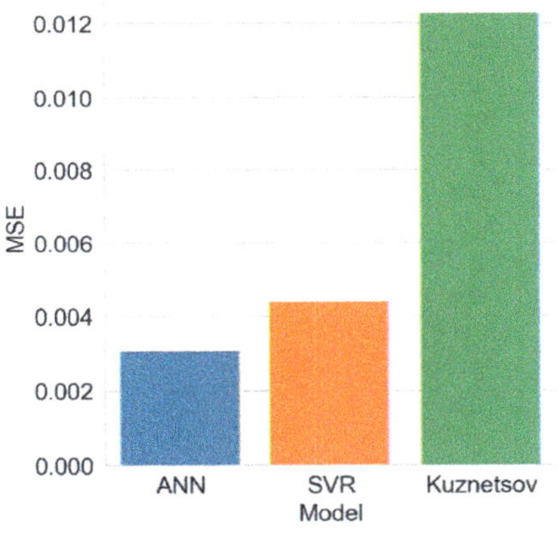

Figure 6. MSE plot for test data.

6. Conclusions and Future Work

The paper successfully demonstrates the potential of achieving higher accuracy in mean rock fragment size prediction using multilayered artificial neural network (ANN) and support vector regression (SVR). Using varied blast data sets from different parts of the world, we obtain training and test sets comprising 90 and 12 blasts, respectively, for building multilayered ANN and SVR models. Both models perform satisfactorily and better than the conventional Kuznetsov empirical model. The paper further demonstrates the possibility to analyze a varied number of hidden layers for a neural network in a less cumbersome way using Keras. Keras makes it less time-consuming to consider the performance of a wide variety of hidden layers and neurons via the Bayesian optimization feature. Thus, multilayered ANN analysis of rock fragmentation, which is typically time-consuming, can be carried out in a relatively shorter time. The end goal here is that blasting engineers would be able to fully exploit the potential of the multilayered ANN architecture for improved performance without having to do manual hyper-parameter tuning. The trained ANN and SVR models could be incorporated into existing fragmentation analysis software to give blasting engineers more accurate options for mean rock fragment size estimation. This incorporation would make it possible for blasting engineers to have access to results from both empirical and machine learning techniques. Blasting engineers would then be able to conduct post-blast analysis to verify the improved accuracy offered by the machine learning techniques. Commercial fragmentation software providers could adopt this integrated approach to gradually build client confidence in the use of machine learning techniques with time.

In the future, we seek to improve model performance via data augmentation. We intend to do this using the variational autoencoding (VAE) technique. VAE is a deep learning technique that fits a probability distribution to a given data set, and then samples from the distribution to create new unseen samples. Thus, the VAE offers a means of augmenting the data set used in this study to improve model training, and thus enhance pattern recognition and prediction. We also seek to build additional rock fragmentation models using other machine learning techniques. The final phase of this project will involve developing robust machine learning-based fragmentation software that will not only predict the mean fragment size but the entire fragment size distribution.

Author Contributions: Conceptualization, R.A., A.J. and S.Z.; Formal analysis, R.A.; Methodology, R.A.; Supervision, A.J. and S.Z.; Visualization, R.A. and A.J.; Writing—Original draft, R.A.; Writing—Review & editing, A.J. and S.Z. All authors have read and agreed to the published version of the manuscript.

Funding: This research received no external funding.

Data Availability Statement: The data presented in this study are available in Hudaverdi et al. [27] (pp. 1322, 1331).

Acknowledgments: We acknowledge the guidance of Ravi Sahu and Oktai Radzhabov of Strayos Inc., Buffalo, NY, USA.

Conflicts of Interest: The authors declare no conflict of interest.

References

1. Rustan, A. (Ed.) *Rock Blasting Terms and Symbols: A Dictionary of Symbols and Terms in Rock Blasting and Related Areas Like Drilling, Mining and Rock Mechanics*; A. A. Balkema: Rotterdam, Holland, 1998.
2. Cunningham, C.V.B. The Kuz-Ram fragmentation model—20 years on. *Bright. Conf. Proc.* **2005**, *4*, 201–210. [CrossRef]
3. Roy, M.P.; Paswan, R.K.; Sarim, M.D.; Kumar, S.; Jha, R.R.; Singh, P.K. Rock fragmentation by blasting: A review. *J. Mines Met. Fuels* **2016**, *64*, 424–431.
4. Adebola, J.M.; Ogbodo, D.A.; Peter, E.O. Rock fragmentation prediction using Kuz-Ram model. *J. Environ. Earth Sci.* **2016**, *6*, 110–115.
5. Kulatilake, P.H.S.W.; Qiong, W.; Hudaverdi, T.; Kuzu, C. Mean particle size prediction in rock blast fragmentation using neural networks. *Eng. Geol.* **2010**, *114*, 298–311. [CrossRef]
6. Jha, A.; Rajagopal, S.; Sahu, R.; Tukkaraja, P. Detection of geological discontinuities using aerial image analysis and machine learning. In Proceedings of the 46th Annual Conference on Explosives and Blasting Technique, Denver, CO, USA, 26–29 January 2020; ISEE: Cleveland, OH, USA, 2020; pp. 1–11.
7. Grosan, C.; Abraham, A. *Intelligent Systems: A Modern Approach*, 1st ed.; Springer: New York, NY, USA, 2011. [CrossRef]
8. Dumakor-Dupey, N.K.; Sampurna, A.; Jha, A. Advances in blast-induced impact prediction—A review of machine learning applications. *Minerals* **2021**, *11*, 601. [CrossRef]
9. Shi, X.Z.; Zhou, J.; Wu, B.B.; Huang, D.; Wei, W. Support vector machines approach to mean particle size of rock fragmentation due to bench blasting prediction. *Trans. Nonferrous Met. Soc. China* **2012**, *22*, 432–441. [CrossRef]
10. Amoako, R.; Brickey, A. Activity-based respirable dust prediction in underground mines using artificial neural network. In *Mine Ventilation–Proceedings of the 18th North American Mine Ventilation Symposium*; Tukkaraja, P., Ed.; CRC Press: Boca Raton, FL, USA, 2021; pp. 410–418. [CrossRef]
11. Ertel, W. *Introduction to Artificial Intelligence (Undergraduate Topics in Computer Science)*, 2nd ed.; Springer: Berlin/Heidelberg, Germany, 2017.
12. Dixon, D.W.; Ozveren, C.S.; Sapulek, A.T. The application of neural networks to underground methane prediction. In Proceedings of the 7th US Mine Ventilation Symposium, Lexington, KY, USA, 5–7 June 1995; pp. 49–54.
13. Krose, B.; van der Smagt, P. *An Introduction to Neural Networks*, 8th ed.; University of Amsterdam: Amsterdam, The Netherlands, 1996.
14. Zhou, J.; Li, X.; Shi, X. Long-term prediction model of rockburst in underground openings using heuristic algorithms and support vector machines. *Saf. Sci.* **2012**, *50*, 629–644. [CrossRef]

15. Shuvo, M.M.H.; Ahmed, N.; Nouduri, K.; Palaniappan, K. A Hybrid approach for human activity recognition with support vector machine and 1D convolutional neural network. In Proceedings of the 2020 IEEE Applied Imagery Pattern Recognition Workshop (AIPR), Washington, DC, USA, 13–15 October 2020; IEEE: Piscataway, NJ, USA, 2020; pp. 1–5.
16. Hamed, Y.; Ibrahim Alzahrani, A.; Shafie, A.; Mustaffa, Z.; Che Ismail, M.; Kok Eng, K. Two steps hybrid calibration algorithm of support vector regression and K-nearest neighbors. *Alex. Eng. J.* **2020**, *59*, 1181–1190. [CrossRef]
17. Nandi, S.; Badhe, Y.; Lonari, J.; Sridevi, U.; Rao, B.S.; Tambe, S.S. Hybrid process modeling and optimization strategies integrating neural networks/support vector regression and genetic algorithms: Study of benzene isopropylation on Hbeta catalyst. *Chem. Eng. J.* **2004**, *97*, 115–129. [CrossRef]
18. Gheibie, S.; Aghababaei, H.; Hoseinie, S.H.; Pourrahimian, Y. Modified Kuz-Ram fragmentation model and its use at the Sungun Copper Mine. *Int. J. Rock Mech. Min. Sci.* **2009**, *46*, 967–973. [CrossRef]
19. Kuznetsov, V.M. The mean diameter of the fragments formed by blasting rock. *Sov. Min. Sci.* **1973**, *9*, 144–148. [CrossRef]
20. Cunningham, C.V.B. The Kuz–Ram model for prediction of fragmentation from blasting. In *Proceedings of the First International Symposium on Rock Fragmentation by Blasting; Luleå, Sweden, 23–26 August 1983*; Holmberg, R., Rustan, A., Eds.; Lulea University of Technology: Luleå, Sweden, 1983; pp. 439–454.
21. Cunningham, C.V.B. Fragmentation estimations and the Kuz–Ram model—Four years on. In Proceedings of the Second International Symposium on Rock Fragmentation by Blasting, Keystone, CO, USA, 23–26 August 1987; pp. 475–487.
22. Rosin, P.; Rammler, E. The laws governing the fineness of powdered coal. *J. Inst. Fuel* **1933**, *7*, 29–36.
23. Clark, G.B. *Principles of Rock Fragmentation*; John Wiley & Sons: New York, NY, USA, 1987.
24. Kanchibotla, S.S.; Valery, W.; Morrell, S. Modelling fines in blast fragmentation and its impact on crushing and grinding. In *Explo '99–A Conference on Rock Breaking*; The Australasian Institute of Mining and Metallurgy: Kalgoorlie, Australia, 1999; pp. 137–144.
25. Djordjevic, N. Two-component model of blast fragmentation. In Proceedings of the 6th International Symposium on Rock Fragmentation by Blasting, Johannesburg, South Africa, 8–12 August 1999; pp. 213–219.
26. Ouchterlony, F. The Swebrec© function: Linking fragmentation by blasting and crushing. *Inst. Min. Metall. Trans. Sect. A Min. Technol.* **2005**, *114*, 29–44. [CrossRef]
27. Hudaverdi, T.; Kulatilake, P.H.S.W.; Kuzu, C. Prediction of blast fragmentation using multivariate analysis procedures. *Int. J. Numer. Anal. Methods Geomech.* **2010**, *35*, 1318–1333. [CrossRef]
28. Ouchterlony, F.; Niklasson, B.; Abrahamsson, S. Fragmentation monitoring of production blasts at Mrica. In Proceedings of the International Symposium on Rock Fragmentation by Blasting, Brisbane, Australia, 26–31 August 1990; McKenzie, C., Ed.; Australasian Institute of Mining and Metallurgy: Carlton, Australia, 1990; pp. 283–289.
29. Hamdi, E.; Du Mouza, J.; Fleurisson, J.A. Evaluation of the part of blasting energy used for rock mass fragmentation. *Fragblast* **2001**, *5*, 180–193. [CrossRef]
30. Aler, J.; Du Mouza, J.; Arnould, M. Measurement of the fragmentation efficiency of rock mass blasting and its mining applications. *Int. J. Rock Mech. Min. Sci. Geomech.* **1996**, *33*, 125–139. [CrossRef]
31. Hudaverdi, T. The Investigation of the Optimum Parameters in Large Scale Blasting at KBI Black Sea Copper Works—Murgul Open-Pit Mine. Master's Thesis, Istanbul Technical University, Istanbul, Turkey, 2004.
32. Ozcelik, Y. Effect of discontinuities on fragment size distribution in open-pit blasting—A case study. *Trans. Inst. Min. Metall. Sect. A Min. Ind.* **1998**, *108*, 146–150.
33. Jhanwar, J.C.; Jethwa, J.L.; Reddy, A.H. Influence of air-deck blasting on fragmentation in jointed rocks in an open-pit manganese mine. *Eng. Geol.* **2000**, *57*, 13–29. [CrossRef]
34. Pedregosa, F.; Varoquaux, G.; Gramfort, A.; Michel, V.; Thirion, B.; Grisel, O.; Blondel, M.; Prettenhofer, P.; Weiss, R.; Dubourg, V.; et al. Scikit-learn: Machine learning in python. *J. Mach. Learn. Res.* **2011**, *12*, 2825–2830. [CrossRef]
35. Chollet, F. Keras. Github Repos. 2015. Available online: https://github.com/fchollet/keras (accessed on 15 June 2020).

mining

MDPI

Article

Development of a Smart Computational Tool for the Evaluation of Co- and By-Products in Mining Projects Using Chovdar Gold Ore Deposit in Azerbaijan as a Case Study

Anvar Mammadli [1], George Barakos [2,*], Md Ariful Islam [3], Helmut Mischo [3] and Michael Hitch [2]

[1] Horizon Education Agency, Baku 1065, Azerbaijan; a.mammadli@hea-eu.com
[2] Western Australian School of Mines: Minerals, Energy and Chemical Engineering, Curtin University, Bentley, WA 6102, Australia; michael.hitch@curtin.edu.au
[3] Institute of Mining and Special Civil Engineering, TU Bergakademie Freiberg, 09599 Freiberg, Germany; md-ariful.islam@mabb.tu-freiberg.de (M.A.I.); helmut.mischo@mabb.tu-freiberg.de (H.M.)
* Correspondence: george.barakos@curtin.edu.au

Abstract: Despite their significance in numerous applications, many critical minerals and metals are still considered minor. Since most of them are not found alone in mineral deposits, their co- or by-production depends on the production of base metals and other major commodities. In many cases, the concentration of the minor metals is low enough not to be considered part of the production. Hence, their supply is not always secured, their availability decreases, and their criticality increases. Many researchers have addressed this issue, but no one has set actual impact factors other than economic ones that should determine the production of these minor commodities. This study identified several parameters, the number and diversity of which gave birth to developing a computational tool using a multi-criteria-decision analysis model based on the Analytical Hierarchical Process (AHP) and Python. This unprecedented methodology was applied to evaluate the production status of different commodities in a polymetallic deposit located in Chovdar, Azerbaijan. The evaluation outcomes indicated in quantifiable terms the production potentials for several commodities in the deposit and justified the great perspectives of this tool to evaluate all kinds of polymetallic deposits concerning the co- and by-production of several minor critical raw materials.

Keywords: multi-criteria decision analysis; AHP; Python; minor critical metals; mining co- and by-products

Citation: Mammadli, A.; Barakos, G.; Islam, M.A.; Mischo, H.; Hitch, M. Development of a Smart Computational Tool for the Evaluation of Co- and By-Products in Mining Projects Using Chovdar Gold Ore Deposit in Azerbaijan as a Case Study. *Mining* **2022**, 2, 487–510. https://doi.org/10.3390/mining2030026

Academic Editors: Juan M Menéndez-Aguado, Oscar Jaime Restrepo Baena and Jessica M. Smith

Received: 28 June 2022
Accepted: 28 July 2022
Published: 30 July 2022

1. Introduction

From the Ages of Antiquity to the present day, humanity has exploited minerals and metals found on the Earth's crust. Prehistoric man is known to have used only a handful of metals including copper, iron, gold, silver, tin, and lead. Thousands of years later, the Industrial Revolution heralded an unprecedented age of rapid industrial and economic growth that was substantially driven by the exploitation of many more minerals and metals. Undoubtedly, the evolution of the modern world has played a significant role in the constantly increasing production of many more minerals and metals used to perform specialized functions or have found applicability in several new applications [1,2].

However, only few metals such as copper, tin, lead, and iron can be found in relatively high concentrations worldwide and are produced in relatively high volumes [3]. From a geological point of view, these metals can either be found alone or mostly as hosts in polymetallic geological formations. Unlike these "major" metals, there are "minor" metals occurring in polymetallic deposits in concentrations sometimes low enough not to be considered feasibly exploitable on their own [2,4]. These metals are deeply embedded in our high-tech products and despite their increasing demand, they are produced in relatively low volumes. In fact, several of these critical metals are recovered only as by-products from a limited number of geographically concentrated ore deposits, thus making their markets dependent on geopolitical strategies and raising concerns regarding their supply [2].

Several researchers have investigated this matter in detail. In 1979, Skinner [4] was one of the first to talk about the sustainable supply of minor metals and referred to possible resource limitations in the future. A few years later, Campbell [5] presented short-run supply curves for primary and secondary metals, indicating the individual behavior and the interconnection between primary, co-, and by-products in terms of their connected supply and the impact this has on their prices. Wellmer et al. [6] justified Campbell's theory and mentioned that many metals are produced exclusively as by-products of other minerals and metals, meaning that their production is strictly limited by the production of the "host" materials to which they are associated.

In recent years, research has intensified, given that new uses and applications for many more minerals and metals have been developed. Verhoef et al. [7] introduced a system of linked cycles in the form of a metals' wheel, showing metal linkages in natural resource processing while illustrating the capacity of available metallurgical processes dealing with impurities in their primary or secondary feed. Reuter et al. [8] introduced a different metal wheel showing the complex interactions between different metals and the economic and thermodynamic recoverability of (co-)elements. Buchert et al. [9] introduced a group of "green minor metals", emphasizing their significant applicability in renewable energy resourcing, and how some minor critical metals are dependent on the mining development of major metals. Willis et al. [10] conducted research on critical by-products of copper, lead, zinc, and nickel with relatively small volumes of production.

Wellmer and Hagelüken [11] published their work related to the security of supply of secondary resources under conditions of economic viability and environmental sustainability. They introduced a feedback control cycle of mineral supply. Their "metal wheel" summarizes the standard technologies for the metallurgical treatment of metal associations involving major and minor metals. The concentric rings of the wheel demonstrate the interconnectivity between the main metals as carrier metals and the co- and by-product metals. Inspired by Reuter et al. [8], Nassar et al. [2] introduced a periodic table of companionability and their metal wheel version (Figure 1). In this wheel, the principal host metals form the inner circle, while companion elements appear on the outer circle at distances proportional to the percentage of their primary production.

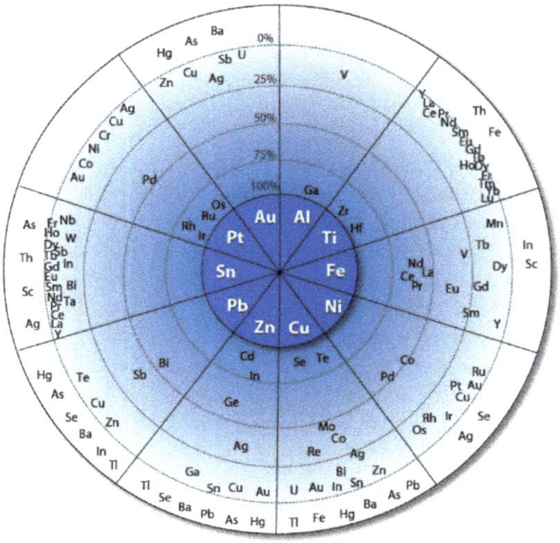

Figure 1. The wheel of metal companionability [2].

Efforts have also been made to quantify the recoverable sources of by-product metals, in the case of cobalt [12] or the case of gallium, germanium, and indium [13,14]. It is worth mentioning that a Minor Metals Trade Association was founded in 1973, when by-product metals were starting to be used in growing mass applications. The Association was formed to guide those involved in the nascent minor metals industry and currently comprises companies from across the globe engaged in all aspects of minor metals activity.

The interconnection of major and minor metals is undeniable, but so is the criticality of several minor metals. However, the concerns about the sustainable supply of critical metals are rarely considered. The selling prices, the additional product extraction costs, and the primary commodity market conditions determine the policies regarding co- and by-products in mining projects. Van Schaik and Reuter [15] tried to associate the sustainability of companion products by linking three core domains: the resource cycle (materials and energy), the natural cycle (society and environment), and the technology cycle (engineering and science). Although this work is mainly related to the recyclability of materials, it is the first effort made to connect the metal wheel with the other cycles. To quantify the recoverable resources of by-product metals, and specifically cobalt, Mudd et al. [12] mentioned a few key parameters to make a realistic estimate of anticipated companion metal availability. The authors focused not only on the economic aspects but also discussed the recovery efficiencies for the minor metals and the benefits of companion metal recovery. Frenzel et al. [13,14] also mentioned the impact of new processing technologies on the by-product production of gallium, germanium, and indium. In other studies, the development of technology has been mentioned to increase by-product recovery efficiency or even reuse waste tailings to recover metals such as rare earth elements [16–18].

Valero et al. [19] adopted the concept presented by Mudd et al. [12] and mentioned that even a single metal may be treated differently within a mining project based on differing grades and quantities. However, their focus was mainly on factors such as the tonnage and commercial prices of commodities. Finally, Renner and Wellmer [20] discussed the impact of volatility drivers on the metal market of both major and minor metals across the globe. According to this study, volatility can result from the fluctuation of commodity prices and political instability, market speculation, and policy responses. Hitch et al. [21] mentioned that the treatment of wastes is not only a matter of new technologies for the feasible recovery of metals but also because of their reactivity characteristics that raise environmental concerns. Thus, waste may turn into a valuable by-product or material for alternative use, while, at the same time, environmental pollution can be prevented.

A series of global supply and demand changes lie at the heart of current developments. Research has shown that the supply of by-product elements is potentially riskier than that of primary elements because the economic health of their associated primary commodity market depends on their recovery [2,22,23]. These changes have led to increased calls for policy responses during the co- and by-production of such products, particularly the most critical ones. For instance, the use of several methods in waste management and the extraction of metals from waste has been expanded in the concept of the circular economy and zero waste production [24,25].

Several mining projects are discussed hereinafter in this study, in which the status of co- and by-products changed during mining operations and production due to several factors. Hence, in addition to costs and revenues, other parameters can also impact the co- and by-production decision before, during, or even after mining operations. However, there is no existing literature that deals directly with this issue. Hence, this work intends to cover this research gap by identifying all possible parameters that can impact the decisions regarding co- and by-product production. In addition to identifying, for the first time, all the impact factors, another objective was to evaluate them in an unprecedented quantitative way overall. This is because the conditions in any mining project vary, and so does the significance of the different parameters. For this reason, a multi-criteria decision analysis technique (MCDA), namely the Analytical Hierarchical Process (AHP) [26], was implemented in the developed methodology, to evaluate all the criteria simultaneously.

Initially, the AHP calculations were undertaken using Microsoft Excel. Nevertheless, the vast number of identified factors and the complexity of the co- and by-production assessment stimulated the development of a new computational tool using Python. The computational power makes this novel tool easier and faster to use while increasing its flexibility by allowing the user to adjust the number of evaluation parameters according to the conditions of any mining project. The final objective of this work was to apply these parameters and their evaluation to a mining case study in Chovdar, Azerbaijan, where a polymetallic deposit based on gold production is exploited. The justification of the methodology through the specific case study denotes the significance of this work.

The purpose of this paper is to present an innovative easy-to-use computational tool that can be applied to any polymetallic project, assist stakeholders to make proper decisions regarding co- and by-production, and thus contribute toward a more secure supply of several critical minor metals and a more sustainable mining industry.

2. Literature Review

A thorough literature review was conducted, regarding the status of primary and companion products, including of waste and tailings in many mining projects that had changed due to various reasons other than economic ones, thus impacting mining plans and production strategies.

2.1. Scandium Production from Red Mud

When the limited availability of a metal is combined with a sudden increase in demand and market volatility, there are several new reasons to proceed with production. This is the case with rare earth elements (REEs) and the extensive publicity they received after the REE crisis of 2011 when fears of supply disruption drove prices up nearly tenfold [27]. The crisis was short-lived, and the prices declined rapidly, but the criticality of REEs remained and, thus, new REE deposits were explored around the world. Parallel to this, several ongoing mining projects, in which rare earths already existed but in small concentrations and were characterized as waste, started investigating their possible production as co- or by-products, including from the waste rock and tailings. For example, rare earth elements, particularly scandium, are occasionally found in bauxite residues, also known as red mud [28,29]. The concentration of rare earths in bauxite residue may vary between 500 and 1700 mg/kg [30]. The increasing importance and the newest developments in processing technology made some mining companies rethink producing REE from waste. For example, a pilot plant is under construction in Greece to investigate the efficiency of leaching and ion exchange on an acid basis to recover scandium from red mud [31,32]. Similar research is being conducted in China [33].

2.2. Borates and Lithium Mining

Similarly, the demand for lithium has increased since its application in batteries has proved highly efficient [34–36]. The exploration boom for battery raw materials included investigations of tailings. Rio Tinto has mined borates in California, US, since 1927 and has recently commenced the production of battery-grade lithium from waste rock at a lithium demonstration plant, being the first top diversified miner to add lithium output to its portfolio, and enhancing the idea of re-evaluating waste rock and tailings [37]. Given the dynamic market of several minor metals, the advanced developments in processing technologies and the need for less waste production, even more producers are reconsidering the possibility of treasures hiding in their tailings. Even tailings from mines having seized operations could also be exploited to recover precious metals, treat the tailings, and mitigate further environmental pollution caused by acid mine drainage. Projects are working toward this direction, such as the Penouta mining project in which tailings are being investigated to recover tantalum and niobium [38], or the Tiouit gold–silver–copper mine in Morocco, where the desulfurization of the old tailings has been investigated [39].

2.3. Mercury Extraction

In modern mining, preserving the environment is considered a top priority. Thus, in addition to the treatment of tailings, dangerous elements such as naturally occurring radioactive materials (NORMs) and toxic elements also receive special treatment. Some of these, such as uranium, thorium, and mercury, are extracted from the waste and treated as by-products even in low non-profitable concentrations. In 1997, Garcia-Guinea and Harffy published a paper in the journal *Nature* with a questioning title about whether mercury mining is undertaken at a profit or a loss [40]. The paper argued how mercury prices have dropped since the 1960s due to many environmental and health problems caused not only by its mining but also by the metal itself. Several publications about mercury pollution [41–43] have built a legacy about how dangerous this element is. Mercury is found mainly in China, Spain, and California, US. The mining district in Almaden, Spain, used to be responsible for 25% of the world's production until operations stopped in 2001 due to the prohibition of mercury mining in Europe [43]. By-product mercury production is expected to continue from large-scale gold–silver mining and processing. There are also reports of small-scale, artisanal mining of mercury in China, Russia (Siberia), Outer Mongolia, Peru, and Mexico [44].

2.4. Marble Quarrying

Primary, co-, and by-products can also be produced from the same commodity but with different quality standards, and different selling prices for different applications. A typical example is steel slag, a by-product of steel making produced during the separation of the molten steel from impurities in steel-making furnaces. Generally, this may not be the case for many metals, but it can be a significant parameter for several industrial minerals, and construction materials in different shapes, sizes, textures, and weights.

Marble, for example, is a dimension stone that is either sold as a whole block or cut into tablets. The size of the block or tablet and the purity of marble are quality standards that affect the price of the final product. Blocks that do not meet the quality standards are crushed, milled, and roasted to become dry pulverized products in different grain sizes. These marble dust and calcium carbonate powders (fillers) are sold for different industrial applications. Dionyssomarble in Attica, Greece, has a long history of exploiting white marble deposits [45]. However, not all products were produced from the beginning. Since 1975, the company has expanded its processing facilities and produced exceptionally clean, aggregate crystalline calcium carbonate powder filler in controlled granular sizes.

2.5. Salt and Potash Rotating Production

Production of some metals such as iron, and some industrial minerals such as salt and potash, can be determined by local demand and supply conditions. Layers of salt and potash follow in geological formations such as bedding planes [46] and can be mined either together or successively. Their co-production flourished in Germany during the 1950s, in the aftermath of World War II, when the reconstruction of the country was at a peak. The German car industry was booming, and the national road network increasingly comprised paved roads. However, the newly paved roads were icy during winter, making driving dangerous. Authorities applied an effective de-icing procedure using salt to clean the roads [47,48].

Production of salt and potash in Germany has focused either on the one commodity or the other, depending on a series of factors that can alter their priority and, in turn, the classification of the two commodities as primary, co-, or by-products. In Sondershausen, Germany, potash production (KCl) started in 1893 and stopped in 1991 (Table 1) due to economic and political reasons (German reunion). However, salt production for de-icing started in the mine in 2004 and continues today [49]. At the Sigmundshall mine, potash production started in 1898, and after 2001, additional production of "Special" potash ($MgSO_4$) took place and expanded the life of the mine (Table 1). The recoverable reserves were depleted in 2018 and the mine finally closed [50]. Furthermore, the Bernburg mine started producing potash in the 1900s and, from 1939, started also producing rock salt. In 1973, the mine stopped producing potash and focused only on salt because a neighboring mine in Zielitz had started potash production in 1969 [51].

Table 1. Potash and salt production history in German mines.

Mine	Potash (KCl)	Potash ($MgSO_4$)	Rock Salt
Sondershausen	1893–1991		2004–today
Sigmundshall	1898–2018	2001–2018	
Bernburg	1900s–1973		1939–today

2.6. Production History of a Silver-Based Polymetallic Deposit

Moving to the eastern part of Germany, we discuss the exploitation of the polymetallic deposit in Freiberg and how different parameters have affected the change in co- and by-products through time. The lead–zinc deposit in Freiberg was discovered in 1168, and it initially attracted interest for its mineralization-bearing silver. Silver was the only mining product until the 18th century. In 1710, the General Melting Administration was founded. Since then, the revenues have also been based on the content of copper and lead [52]. The latter, and the further development of the metallurgical technology, resulted in commencement of the production of lead officially as a by-product in approximately 1820 (Figure 2).

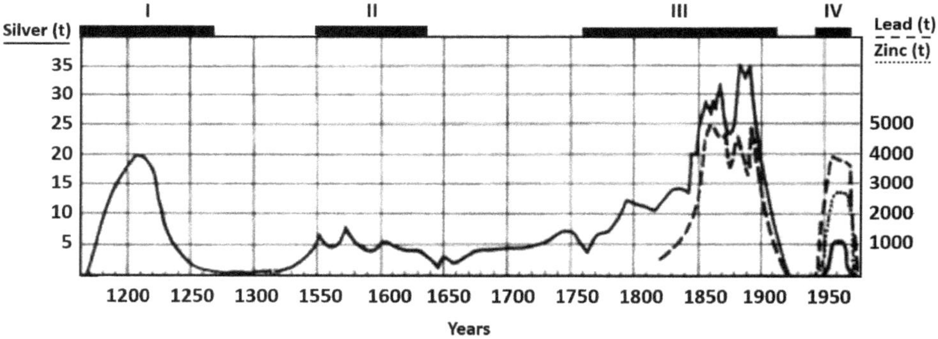

Figure 2. The history of production in Freiberg (modified after Bayer [53]).

The decline of Freiberg's silver mining began with the introduction of gold currency (Goldmark) in the German Empire by law in 1873. The price of silver decreased by half from 1880 to 1898 due to silver deliveries from South America. The prices of by-products lead and zinc decreased massively due to overproduction worldwide. Hence, from 1903–1913 it was decided to shut down all mines [53]. However, due to the preparations in Germany for World War II, there was an increased demand for non-ferrous metals from 1933 onwards. Therefore, from 1935, Freiberg mining resumed. Lead and zinc became the primary commodities for geostrategic reasons, while silver was mined as a by-product. After the end of the war, the mining and metallurgical plants were nationalized in 1961 and closed for economic reasons in 1969 [53].

2.7. Coal and Uranium Rotating Production

Not far from Freiberg, in the Döhlen Basin, coal (initially) and uranium (afterwards) were produced from the same mine site. Mining for hard coal in the area is known to have taken place since the 15th century with rural extraction [54]. Until 1930, small mining companies exploited the area for coal. In 1947, however, SAG/SDAG Wismut started mining uranium bound to hard coal for nuclear armament purposes of the former Soviet Union. Mine ownership alternated between Wismut and the local hard coal mining companies several times. The mining of hard coal for energy purposes stopped in 1967. In 1968 the mine was transferred for one last time to SDAG Wismut. From that time on, until 1989, coal was only mined for its uranium content [54].

3. Materials and Methods

The research methodology developed in this work was initially based on collecting and analyzing information and data from a substantial quantity of literature sources and actual mining projects, some of which have been discussed in the previous sections. What is specific about these mining projects is that, through the years of mining operations and production, the status of their co- and by-products changed due to several factors. These data sets were then used to identify and classify all possible factors that can impact the determination of primary, co-, and by-products in a mining project.

The substantial number and diversity of criteria led to using a multi-criteria-decision-analysis (MCDA) process such as the Analytical Hierarchical Process (AHP) to simultaneously evaluate all the parameters. This MCDA technique has been used in making decisions based on multiple criteria in numerous case studies from a wide range of disciplines.

Depending on the different conditions of any given mining project, these parameters have different levels of importance each time. Therefore, AHP compares the factors and applies weights to them. Accordingly, the multiple final options for each product can also exist. Thus, AHP was further applied to prioritize the final decisions, calculate percentages, and indicate which are preferable on every occasion. As a result, an MCDA tool was developed that can be applied in the evaluation of any polymetallic mining project to determine the main, co-, and by-products. The developed algorithm was computed with the help of Python to create a smart computational tool that will help run the calculations faster and more efficiently.

Data and information from a polymetallic mining project were implemented in the newly developed MCDA tool to test its efficiency. The case study is a gold mining project in Chovdar, Azerbaijan, where gold is the main product, and silver and mercury are produced as by-products. Azergold is the mining company that runs the Chovdar open-pit mining project. The ongoing exploration has revealed additional resources that extend to a substantial depth. For this reason, the company is investigating the possibility of soon transiting to underground mining. Interestingly, in the additional discovered resources, there are a series of other minerals and metals in lower concentrations than those of gold and silver. Accordingly, the developed computational tool was used to determine whether these minerals and metals can be defined as co- or by-products.

4. Setting the Evaluation Parameters

Based on the existing literature, the ore grade and the prices, costs, and reserves determine the revenues of a commodity in a mining project. In some mining projects, when additional resources are found, and the concentrations of some minor elements are significantly increased, this indicates that such elements' status may change from waste to by-product or from by-product to co-product. Parameters other than the price that can interact with the ore grade are the recovery rate during the processing of the ore and the environmental effects if the commodity with the high ore grade happens to be a toxic or radioactive element.

Hence, market, technological, environmental, and socio-political factors were identified, in addition to the apparent economic parameters. The availability of a commodity is an important parameter directly associated with the supply of the commodity in the market, especially for minor metals, and depends on the mining production and processing of the primary commodities. The imbalances between metal supply and demand, actual or anticipated, have inspired the concept of metal criticality [55]. However, the criticality of a commodity is not only dependent on this one parameter. A detailed criticality evaluation includes data from widely varying fields and sources of information, including geology, mineability, technology, the environment, human behavior, the assessment of experts, and many more. Thus, environmental implications, lack of efficient processing techniques and capacity, vulnerability to supply restrictions, and geopolitical issues are some of the most critical factors affecting certain commodities' criticality.

Like the two parameters mentioned above, the market volatility of a commodity is another equally significant factor that can determine the production status of metals. When the limited availability of a commodity is combined with market volatility and a sudden increase in demand, then it seems that there are several new reasons to produce this metal. Finally, another market parameter that needs to be considered is the local demand for commodities and how this can affect their production status.

The local demand is not a factor that applies to most metals traded worldwide. However, it would undoubtedly affect minor minerals and metals that could be produced as by-products, contribute to the revenues, lower the waste production and disposal, and finally meet the demands of local societies. The locality of a mining product is directly related to the extraction and logistic costs, prices, and socio-political factors.

Through technology, the mining industry has overcome many obstacles. New options for increasing productivity are being generated by the evolving technology of the mining industry [56]. A significant technological factor regards the quality standards that a product shall meet to be determined as a primary, co-, or by-product, or waste. It may seem this parameter does not apply to many metals but only to some industrial minerals and dimension stones, as shown in the literature review. However, even regarding the processing and refinement of metal alloys, if the end products do not meet the quality standards of the market or a specific customer, then their value is depreciated, and their feasible production may well be at risk. Low production efficiency can affect extraction costs, not to mention the quality standards. High production efficiency can boost the feasible production of minor and low concentration elements in a deposit and determine them as potential by-products. Most importantly, increased production efficiency offsets declining ore grades and mining cost inflation that threaten the mining industry. In addition, the metal recovery rate, also known as the mineral recovery percentage, indicates the percentage at which valuable metals are expected to be available for sale after the refining process has taken place.

Mining will always impact people and the environment, either positively or negatively. The presence and content of NORMs and other toxic compounds can entail high environmental risks and may require particular attention and close monitoring [57]. Such metals require special treatment either as products or waste, and, even though they are usually found in small concentrations, it is often cheaper to process them as by-products rather than treat them as waste. Another group of potential contaminators is that of the greenhouse gases responsible for the greenhouse effect primarily associated with coal mining [58,59]. An additional factor interconnecting with the presence of NORMs and toxic compounds from the extraction of minerals and metals regards the treatment and disposal of wastes and tailings [60].

Evaluation is also needed of the "mining friendliness" of the commodities produced in a mine. Not all commodities are easy and environmentally friendly to extract. Some elements have gained a reputation for being extremely hazardous when mined. Even when the actual risk of contamination is low due to insignificant concentrations or when the actual contamination is minimized due to sufficient safety measures, opposition to mining-specific commodities can be substantial. In fact, the mining industry considers the Social License to Operate as the most important business risk to be revoked by local communities if unsatisfactory conditions occur [61,62].

Therefore, the social acceptance of extracting specific commodities in a mine is an essential factor. It is also significant to evaluate the legislation status that governs the mining industry in a country and the specific legislation acts that may support or prohibit the production of specific commodities. Finally, the strategic importance of specific commodities is an important factor that should never be neglected. The classification of a metal as strategic and critical not only for economic but also for political and strategic reasons may influence its production status from waste to a by-product or even co-product. The strategic importance of a commodity can affect the criticality, availability, and volatility of its market, not to mention its price. It can also affect the social acceptance and amendment of legislation related to its production.

Accordingly, 18 qualitative and quantitative parameters were determined and classified into five categories according to the relevance of the criteria in the respective categories (Figure 3). Many of these parameters have never been considered before, and no similar classification has been introduced in the literature. Some of the parameters may overlap with others. At the same time, factors can be attributed to more than one of the main categories in which they are classified in their simultaneous evaluation. The clustered criteria are structured in such a way that, in each category, they do not exceed the number of 7 ± 2 because of the general limitations of the human mind, which is capable of handling only so many conceptual objects and discrete figures at a time [63,64]. Hence, criteria belonging to the same category can be easily evaluated and compared on a pair-wise basis.

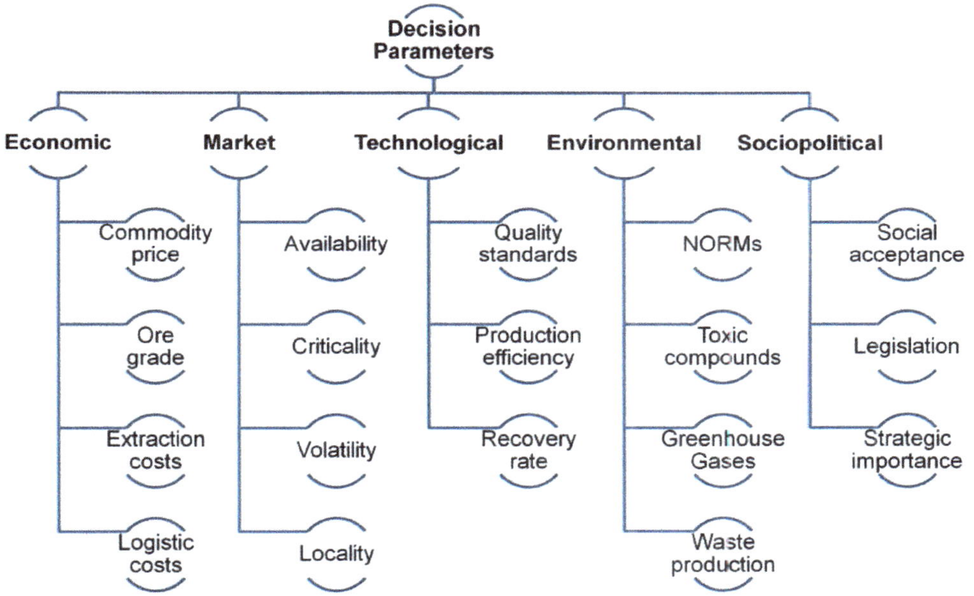

Figure 3. Classification of the parameters for the determination of co- and by-products production.

The overall classification is hierarchical so that all criteria are rightfully prioritized. Nevertheless, depending on the deposit properties and the conditions of the examined mining project, not all criteria need to be evaluated in every case study. When a specific parameter is neutral or does not affect the product status determination, it can be excluded from the evaluation.

5. Development of the Decision Tool

Whether a mining product is characterized as primary, secondary, or waste based on so many factors is a sophisticated process. Such a complicated problem needs to be decomposed into simple assessments without neglecting that some elements have a more significant impact on the decision making than others.

Decision making, in general, is explained as a selection process in which the best alternative is chosen from alternative sets to reach an aim or multiple aims. The process alone is not concerned with defining the objectives, designing specific alternatives, or evaluating consequences; decision making offers simple techniques and procedures to reveal preferences and choices in multivariable problems. Such techniques are described as multi-criteria decision analysis (MCDA) or multiple-attribute decision making (MADM). These techniques solve problems in which discrete alternatives can be selected from a finite set [65,66]. Existing MCDA methods include value measurement models, such as the Analytical Hierarchical Process (AHP) and Multiple-Attribute-Utility Technique (MAUT); goal-, aspiration-, or reference-level models, such as the Technique for Order Preference by Similarity to Ideal Solution (TOPSIS); and outranking models, such as the Elimination and Choice Translating the Reality (ELECTRE) and Preference Ranking Organization Method for Enrichment Evaluation (PROMETHEE) methods [65]. Each method has its strengths and weaknesses in different areas, and it is difficult to say one is better than another; ultimately, it depends on the specific problem that needs to be solved.

Considering the nature of the decision-making problem in this work, AHP was selected over the other MCDA methods. This technique is preferred for its ability to rank alternatives in order of their effectiveness when conflicting objectives or criteria must be satisfied [67,68]. Furthermore, AHP can detect inconsistent judgements and estimate their degree of inconsistency [67]. Moreover, the parameters determined in the previous section are classified into separate categories, making AHP the ideal decision-making method to decompose the problem and build hierarchies of the individual criteria. Finally, the AHP preferences and pair comparisons can be easily computed.

Initially, the algorithm was developed in Microsoft Excel and later in the form of a Python computational tool to make the calculations faster and more efficient. The following sections discuss the development of the algorithm and the computational tool.

5.1. Development of the Product Decision Tool

The decision-making algorithm is mapped into a generic AHP hierarchy (Figure 4), in the order of the five categories, the 18 criteria, and the three product options. To facilitate easier data manipulation in the evaluation process, the categories, criteria, and options were coded (Tables 2–4).

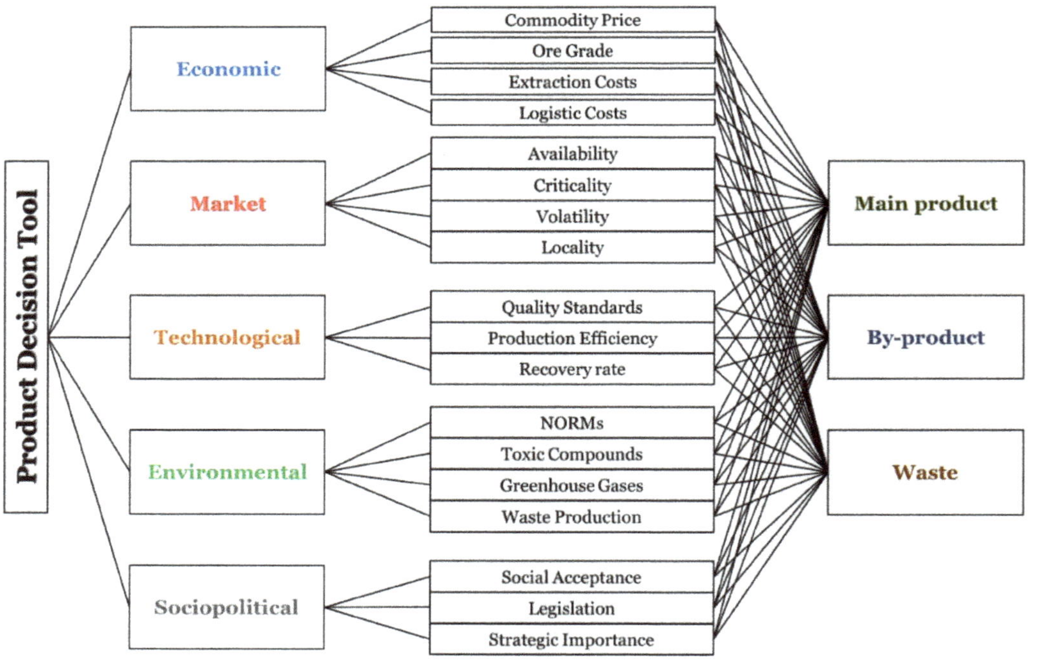

Figure 4. Hierarchy structure of the Product Decision Tool.

Table 2. Category names and codes.

Codes	Categories
ECO	Economic
MAR	Market
TEC	Technological
ENV	Environmental
SOC	Sociopolitical

Table 3. Criteria names and codes.

Codes	Criteria
C_1	Commodity price
C_2	Ore grade
C_3	Extraction costs
C_4	Logistic costs
C_5	Availability
C_6	Criticality
C_7	Volatility
C_8	Locality
C_9	Quality standards
C_{10}	Production efficiency
C_{11}	Recovery rate
C_{12}	NORMs
C_{13}	Toxic compounds
C_{14}	Greenhouse gasses
C_{15}	Waste production
C_{16}	Social acceptance
C_{17}	Legislation
C_{18}	Strategic importance

Table 4. Codes for the product options.

Codes	Categories
PRPRO	Primary product
BYPRO	By-product
WASTE	Waste

The purpose of AHP is to assist decision makers in organizing their judgements to make more effective product decisions by bringing the evaluation to the level of pair-wise comparisons of components with respect to attributes and alternatives. The AHP method uses both qualitative and quantitative variables, and it is not only useful for making decisions, but also for prioritizing tangible and intangible criteria by setting weight factors on them. To make these comparisons, a fundamental scale introduced by Saaty [26] is used to indicate how many times more important one element is over another (Table 5). The result of the pairwise comparisons over n criteria is summarized in a $n \times n$ reciprocal matrix (Table 6), where elements represent the pair-wise comparisons. Each entry of the matrix represents the importance of one criterion relative to the other.

The next step is to compute the vector of weights based on the theory of eigenvector procedure in two steps. First, the matrix is normalized, and the criteria weight vector is then built. The sum of all elements in the weight vector is equal to 1 and shows the relative weights among the compared criteria. Since the comparison is based on subjective evaluations, the consistency of the comparisons is checked using a consistency index. If the degree of inconsistency in judgements is acceptable, the efficiencies of all alternatives on a criterion are normalized to eliminate the effect of different units of measure. The matrix of the normalized efficiency outcomes is finally multiplied by the eigenvector to obtain the aggregated AHP priority score. The decision is then made based on the logic that the higher the AHP priority score for an alternative, the more preferable this alternative.

Table 5. The fundamental scale of AHP [26].

Relative Intensity	Definition	Explanation
1	Of equal value	Two elements are of equal value
3	Slightly more value	Experience slightly favors one element over another
5	Essential or strong value	Experience strongly favors one element over another
7	Very strong value	An element is strongly favored, and its dominance is demonstrated in practice
9	Extreme value	The evidence favoring one over another of the highest order of affirmation
2, 4, 6, 8	Intermediate values	When compromise is needed

Table 6. Pair-wise comparison matrix of the main categories.

	ECO	MAR	TEC	ENV	SOC	Weights
ECO	1	$a_{eco,mar}$	$a_{eco,tec}$	$a_{eco,env}$	$a_{eco,soc}$	w_{eco}
MAR	$1/a_{eco,mar}$	1	$a_{mar,tec}$	$a_{mar,env}$	$a_{mar,soc}$	w_{mar}
TEC	$1/a_{eco,tec}$	$1/a_{mar,tec}$	1	$a_{tec,env}$	$a_{tec,soc}$	w_{tec}
ENV	$1/a_{eco,env}$	$1/a_{mar,env}$	$1/a_{tec,env}$	1	$a_{env,soc}$	w_{env}
SOC	$1/a_{eco,soc}$	$1/a_{mar,soc}$	$1/a_{tec,soc}$	$1/a_{env,soc}$	1	w_{soc}

To ease the assessment process, pair-wise comparisons of the criteria are separately undertaken for each category. A pair-wise comparison of the categories is also undertaken to show their respective relevant importance. Hence, six matrices are generated but, for the sake of space, only the pair-wise comparison of the categories is illustrated in Table 6. This process applies weight factors to all categories and criteria.

Like probabilities, weights are absolute dimensionless numbers between zero and one. Depending on the problem, "weight" can refer to importance, preference, and likelihood, or the decision makers can consider another relevant parameter. Weights are distributed in a hierarchy according to their architecture, and their values depend on the information entered by users of the process [26]. The criteria weights and options are intimately related but need to be considered separately. The priority of the goal and the alternatives always add up to 1 (or 100%). This can become complicated with multiple criteria levels but, if there is only one level, their priorities also add to 1.

Two additional concepts apply when a hierarchy has more than one level of elements, like in this case where we have the categories and the involved parameters: local and global priorities. The local weights here (w_i) represent the relative weights of the nodes within each closed group of siblings (criteria) concerning their parent (category). These local priorities of each group of criteria add up to 1.000 or 100% (Equations (1) and (2)). The global weights (gw_i) are then obtained by multiplying the local weights of the siblings (criteria) by their parent's (category) global priority (Equation (3)). Hence, the global weights for all parameters in the level add up to 1 or 100% (Equation (4)).

$$w_{eco} + w_{mar} + w_{tec} + w_{env} + w_{soc} = 1 \quad (1)$$

$$w_i + w_{i+1} + \ldots + w_n = 1 \quad (2)$$

$$gw_i = w_{zzz} \times w_i \quad (3)$$

$$gw_i + gw_{i+1} + \ldots + gw_t = 1 \quad (4)$$

where:

n is the number of parameters in each category;

zzz represents each of the five categories (ECO, MAR, TEC, ENV, and SOC);

t is the total number of criteria (in this case t = 18).

The next step is to compare all three options (primary or co-product, by-product, and waste) per criterion. This process will generate 18 3 × 3 matrices, the general version of which is illustrated in Table 7. The priorities (w_{yCi}) for the (y = 3) options are calculated with the same procedure as for the categories and criteria. Consequently, the weights in each matrix calculation add to 1.

Table 7. Pair-wise comparison matrix of the options with respect to criterion Ci.

	PRPRO	BYPRO	WASTE	Weights
PRPRO	1	$a_{prpro,bypro}$	$a_{prpro,waste}$	w_{1Ci}
BYPRO	$1/a_{prpro,bypro}$	1	$a_{bypro,waste}$	w_{2Ci}
WASTE	$1/a_{prpro,waste}$	$1/a_{bypro,waste}$	1	w_{3Ci}

Each weight (w_{yCi}) is multiplied by the global weight (gw_i) of the respective criterion and summed to the score for each option (Equation (5)).

$$OPTION_y = (gw_i \times w_{yCi}) + (gw_{i+1} \times w_{yCi+1}) + \ldots + (gw_t \times w_{yCt}) = 1 \quad (5)$$

The outcome for all preferences indicates which product option is the most suitable. The sum of all options is equal to 1 or 100%. The stronger an option, the more apparent the decision that needs to be made. However, when two options are close to each other, more detailed evaluations may need to be made.

The procedure is separately conducted for each mineral or metal. It needs to be individually repeated for all minerals and metals, indicating whether each should be considered a primary product, co-product, or by-product, or be treated as waste. The criteria and the options are evaluated (comparing pairs) without neglecting the priority and importance of any categories or parameters. The classification is done according to the relevance of the criteria in the respective categories, even though some parameters could be included in other categories, and several criteria are interconnected. Depending on the properties of the element under evaluation and the conditions of the examined mining project, not all criteria need to be evaluated. When a specific parameter is neutral or does not affect the product selection, it can be excluded from the evaluation.

This was an issue during the initial development of the algorithm in Microsoft Excel. Changing the tool's structure by adding or excluding criteria to meet the conditions of each element or project under examination required time and effort. This problem was solved with the development of the Python computational tool, which will be discussed in the following sections. Another solution was to modify Saaty's fundamental scale (Table 4). While keeping the general structure of the scale the same, the assigned relative densities of unusable parameters can be rearranged to acquire the lowest possible weights. The same adjustments can be made during the evaluation of the options The consistency of the calculations can be checked again after the rearrangements.

5.2. Development of the Computational Tool

The next step was to convert the developed algorithm to Python Code and use Tkinter, which is the standard graphical user interface (GUI) library for Python, to build an easy-to-use and fast-calculating computational tool. The developed tool uses three types of input data: general input data, data based on the number of categories and criteria, and data based on the number of options. The categories and options numbers are the primary input values (Figure 5); other input variables, including the names of categories, number of criteria, and names of options, are dependent on the primary input values (Figure 6a,b). Next is the comparison of the categories and criteria in pairs in the generated n × n matrices. The reciprocity of the matrices allows for the automatic generation of half the inputs. In addition, the user can insert fractions when the comparisons favor the second parameter over the first (Figure 7). In AHP, pairwise comparisons can be made by more than one decision maker, and a geometric mean can be used to consider all the options.

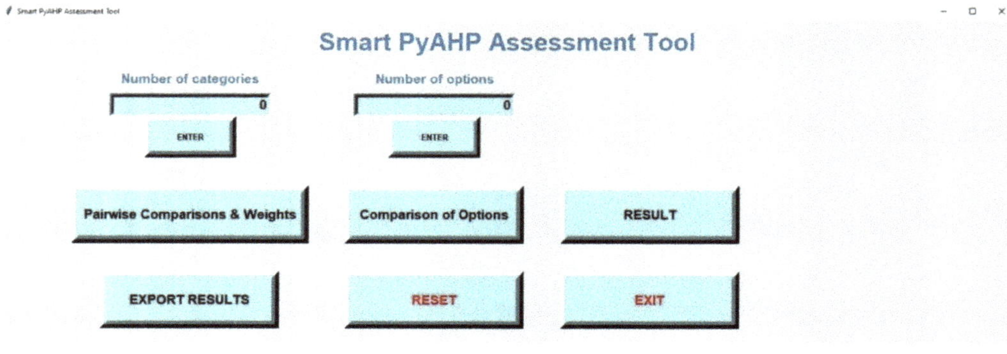

Figure 5. Main window of the computational tool.

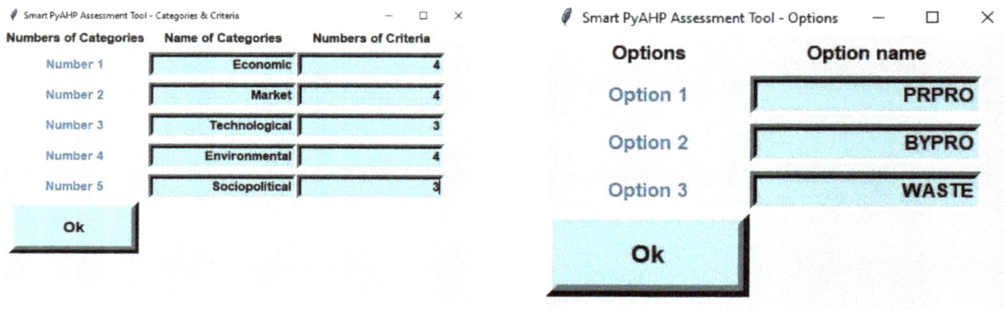

(a) (b)

Figure 6. Insertion of: (**a**) the names of categories and numbers of criteria; (**b**) the names of options.

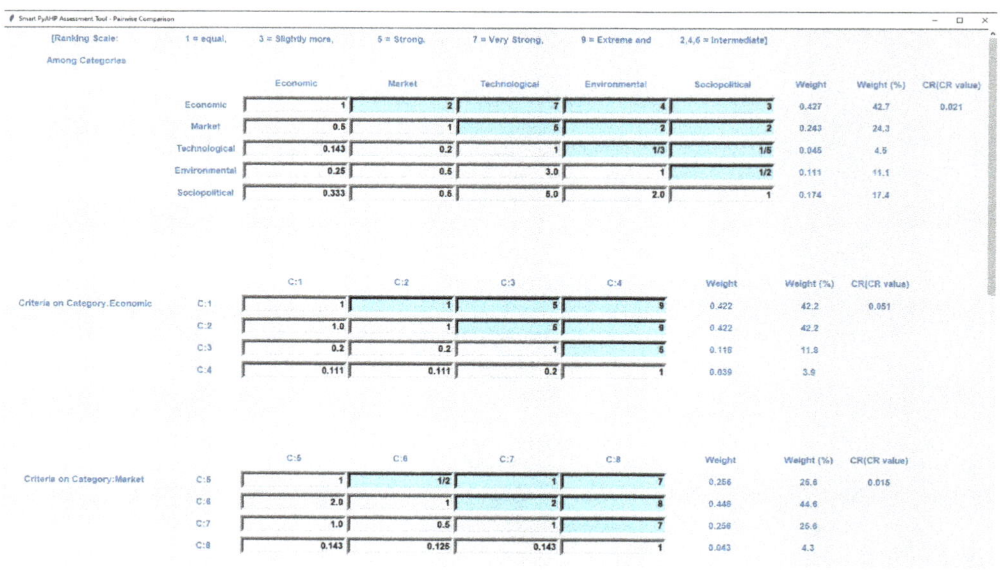

Figure 7. Pair-wise comparisons inserted in the generated matrices.

The tool generates the local and global weights for all criteria (Figure 8) and checks the consistency of the comparisons. Separate calculations are made to evaluate the three options for each weighted criterion. The overall process output is given in percentages of preference for each of the three options (Figure 9). Hence, the user can identify the preferences for the mineral or metal under investigation as a potential primary, co-product, or by-product, or if it shall be treated as waste.

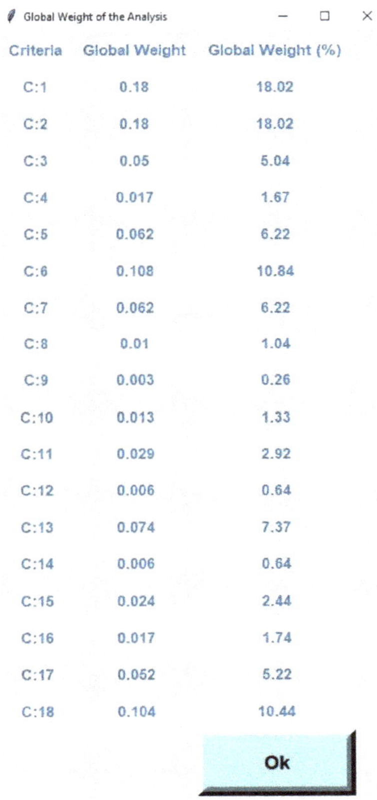

Global Weight of the Analysis

Criteria	Global Weight	Global Weight (%)
C:1	0.18	18.02
C:2	0.18	18.02
C:3	0.05	5.04
C:4	0.017	1.67
C:5	0.062	6.22
C:6	0.108	10.84
C:7	0.062	6.22
C:8	0.01	1.04
C:9	0.003	0.26
C:10	0.013	1.33
C:11	0.029	2.92
C:12	0.006	0.64
C:13	0.074	7.37
C:14	0.006	0.64
C:15	0.024	2.44
C:16	0.017	1.74
C:17	0.052	5.22
C:18	0.104	10.44

Ok

Figure 8. Generation of the global weights for all criteria.

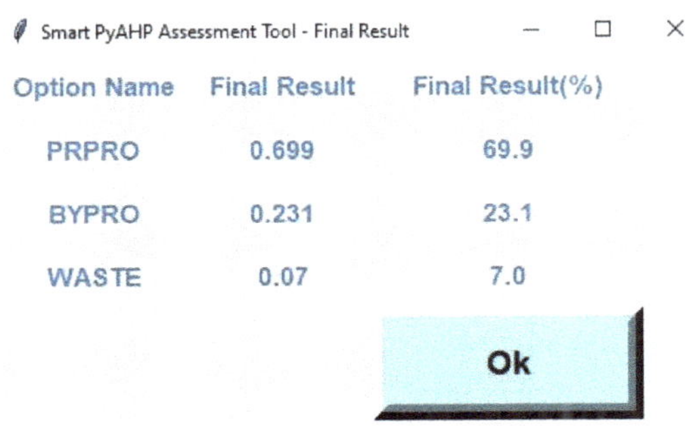

Smart PyAHP Assessment Tool - Final Result

Option Name	Final Result	Final Result(%)
PRPRO	0.699	69.9
BYPRO	0.231	23.1
WASTE	0.07	7.0

Ok

Figure 9. Generation of the preferences for all three options.

The computation of an AHP algorithm in Python is a sophisticated process carried out under various functions' directions (Figure 10a). Several lists and dictionaries were required to overcome the calculations' complexity, considering that local and global weights had to be generated and consistency had to be checked (Figure 10b).

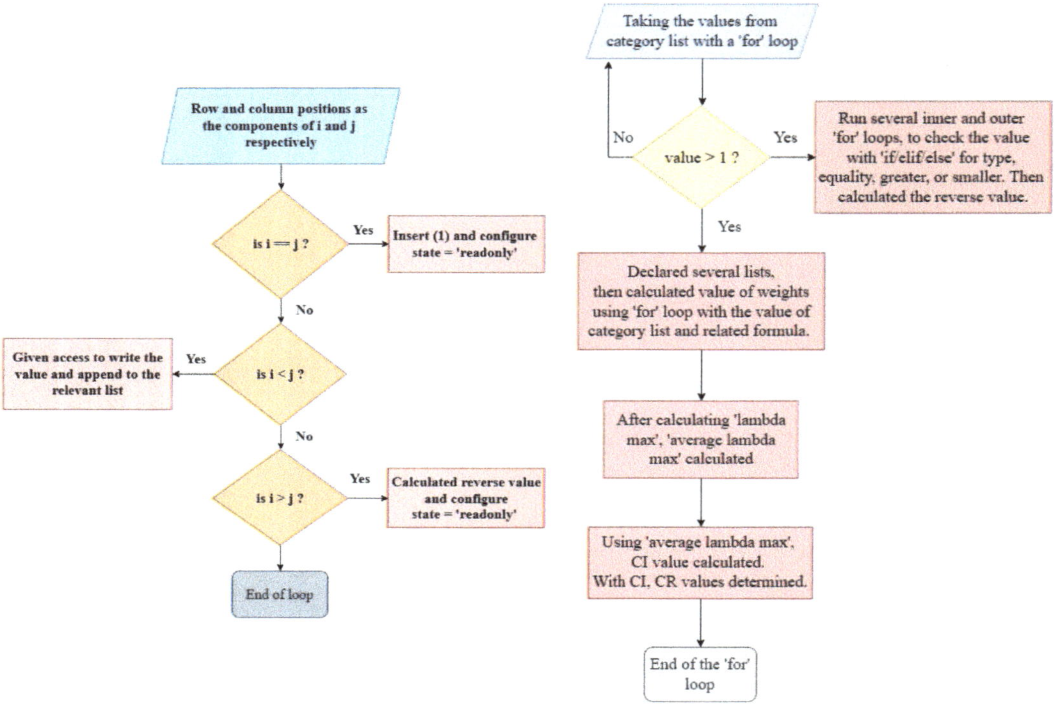

(**a**) (**b**)

Figure 10. Flowcharts for: (**a**) generating matrices to input and collect values; (**b**) the calculation of weights and consistency indexes.

The outcomes of all calculations can be exported in CSV files and further processed in Microsoft Excel. The next step toward optimizing the developed tool is to allow the user to insert data from CSV or ASCII files.

6. Results and Discussion

Following the theoretical development of the AHP decision tool, the assessment of one case study is described in this work. The Chovdar gold mining project in Azerbaijan was selected for the application of the computational tool. Gold is the main product of this mine, and silver is a by-product. However, detailed exploration activities have revealed the presence of other minerals and metals in smaller concentrations.

The exploitation is scheduled in two phases; the first phase has already started, and surface mining is applied, while a feasibility study is also being prepared for the second phase, in which exploitation will transition to underground mining operations. The concentrations of all metals other than gold and silver are insignificant during the first phase. However, in the second phase, the resources to be mined include higher concentrations of metals such as copper, iron, and bauxite. It has not been clarified whether the mining company—Azergold—will exploit these additional elements as co- or by-products. Hence, applying the developed tool in this case study may significantly contribute to the actual decision making.

All authors' calculations and assessments were made together and are based on publicly available data and information provided by the managers, engineers, and personnel of Azergold during a three-month internship (September–November 2019) of the first author at the mine site in Chovdar, Azerbaijan. Feasibility Study and Environmental Impact Assessment reports for the deposit are pending and, thus, not enough technical and economic data are available for a more precise assessment of the potential products. Nevertheless, existing data can yield a first good estimation for all commodities.

6.1. The Chovdar Polymetallic Deposit in Azerbaijan

Chovdar is known as a sizeable gold-sulfide deposit discovered relatively recently (in 1998) and run by Azergold in an area known for its several gold–silver–copper-low-sulfide occurrences and mineralization points. It is in western Azerbaijan's northern part, approximately 45 km west of Ganja and 370 km west of Baku [69]. The main exploration activities lasted until 2011, and mining operations commenced in 2012.

Two natural types of ores have been established in the Chovdar gold ore deposit: oxidized and primary sulfides distinguished by mixed semi-oxidized ores (Figure 11). The oxide mineralization constitutes the upper section of the breccia deposit and varies in thickness from 60 to 80 m. Below the weathered material, the thickness of the primary sulfide mineralization ranges from 100 to 200 m, and extends to about 250 m below the surface.

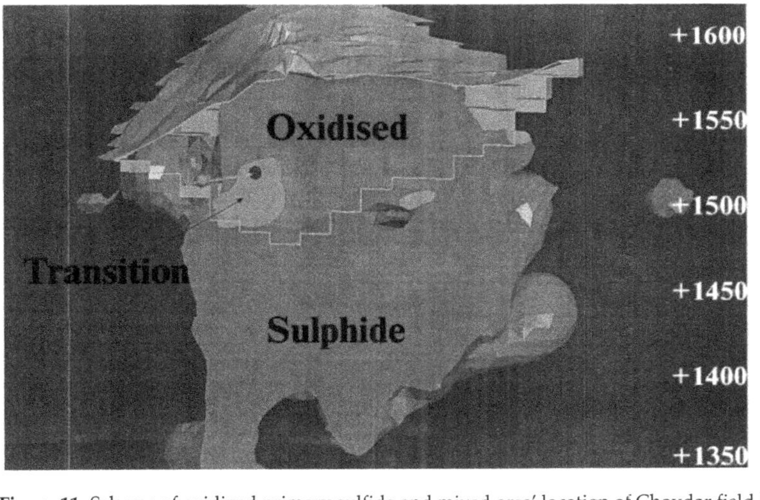

Figure 11. Scheme of oxidized primary sulfide and mixed ores' location of Chovdar field [70].

The indicated and inferred mineral resources for the oxidized part of the deposit are estimated at 4.4 Mt. For the sulfide phase of the deposit, the resource estimate is 13.7 Mt [69,70]. The cut-off-grade for the mineral resource reporting is set at 0.5 gr/tonne of gold. The exploration results resulted in exploiting the mineralization in two phases. Phase One is to exploit the oxidized mineral reserves from an optimized open pit, and Phase Two is to develop an underground mine to subsequently exploit the remaining oxidized mineralization and sulfide mineralization [69].

The Chovdar process plant is located approximately one kilometer south of the open pit. It comprises the entire treatment process from ore size reduction, beneficiation, heap leaching, carbon processing, electro-winning, refining, cyanide recovery, copper recovery, and, finally, cyanide destruction before tailings discharge [69]. The end-product is in the form of gold–silver alloys, shipped to Switzerland for further processing [71]. Although present in low concentrations, mercury cannot be disposed of as waste on-site; thus, it is also shipped off-site. Further mercury data are unavailable due to confidentiality restrictions of the company.

Exploration continues through the strategic phases of thorough assessment and evaluation during the life of a mine. Azergold has started geophysical and drilling operations to increase the reserves on the near and far flanks of the deposit. An interesting piece of information to note while transitioning from the oxide to the sulfide phase is the changing of concentration percentages in several metals found in the mineralization of Chovdar. Gold is the main product, silver is mined as a by-product, and mercury is extracted as waste. Various metals such as copper, iron, zinc, and aluminum, among others, are found in relatively low and uneconomic concentrations [69,71].

However, in the sulfide phase, the concentration of some minor metals is increasing to be significant enough, and should attract the attention of the project managers and make them reconsider their production. A detailed analysis was carried out for the chemical composition of the elements based on samples. The gold content ranges from 0.65 to 3.85 ppm, and is the leading commercially valuable component in all considered samples. Silver, having content ranging from 2.5 to 23.2 ppm, is particularly interesting for the following extraction. Moreover, a relatively high content of copper (0.825%) has been revealed in some samples. This suggests the possibility of the subsequent efficient extraction of the metal from the primary-sulfide ore types of the deposit [70]. Similarly, increased grades of iron and aluminum indicate that further techno-economic analysis should be undertaken for the potential production of these elements. Raising iron and aluminum ore grades may not be as economically attractive as for copper, but it certainly should attract the interest of the project managers.

The transition from the oxide to sulfide phase will probably require a significant change in the processing method. Gold particles in the sulfide phase are mostly encapsulated in pyrite and, thus, are not amenable to cyanidation. Pre-oxidation of the pyrite was necessary to liberate gold particles or provide a path for cyanide to contact the gold. This will probably require the process plant to be modified before sulfide exploitation.

6.2. Evaluation of Products in the Chovdar Mining Project

Six potential products were individually evaluated: gold, silver, copper, iron, aluminum, and mercury. Based on the history of production during the oxidized phase, gold was the first commodity to be evaluated, followed by silver. Assessments of copper, iron, aluminum, and mercury were then conducted. Once each metal was evaluated, the results were also considered for the following commodities' assessment.

In each evaluation, the first action is to prioritize the categories between them and then separately make cross-comparisons of the criteria in each category. Hence the global weights are generated for all parameters. Then, the three options for each commodity (primary product, by-product, waste) are evaluated for suitability to the respective metal concerning every parameter. Finally, options preferences (in %) are calculated for each commodity. Table 8 discusses the global weights calculated for all potential products.

The evaluation of gold indicates that the economic parameters are considered with the highest priority, and the market and sociopolitical factors follow in percentages. The environmental criteria are relatively less important, and the technological parameters are ranked last. The price and grade of gold in the deposit are the most significant factors, followed by its criticality and strategic importance. The toxic compounds used in the processing (cyanide) also seem to have a remarkable impact on the evaluation.

Table 8. Calculated global criteria for all potential products in the Chovdar deposit.

Criteria	Gold (%)	Silver (%)	Copper (%)	Iron (%)	Aluminum (%)	Mercury (%)
Commodity price	19.3	16.2	12	13.8	13.9	0.5
Ore grade	19.3	16.2	33.8	31.1	31.5	0.5
Extraction costs	4	1.8	5.5	5.7	5.8	2.7
Logistic costs	2.1	1.8	2.6	2.5	2.6	0.5
Availability	5.6	8.7	3.1	3.6	3	1.2
Criticality	10.8	17.3	3.1	3.6	3	1.2
Volatility	5.6	8.7	3.1	3.6	3	1.2
Locality	0.8	1.2	3.1	7.1	6	1.2
Quality standards	0.4	0.4	1.2	1.6	1.7	1.1
Production efficiency	2.4	3	7.5	2.9	3.1	10
Recovery rate	3.4	3.8	11.2	5.2	5.6	10
NORMs	0.7	0.6	0.5	1.9	2.1	2.4
Toxic compounds	6.6	5.2	2	1.9	2.1	21.6
Greenhouse gasses	0.7	0.6	0.5	1.9	2.1	2.4
Waste production	3.5	2.7	3.8	3.9	4.1	14.4
Social acceptance	1.5	1.2	2.3	3.2	3.5	9.3
Legislation	4.5	6	2.3	3.2	3.5	18.8
Strategic importance	8.9	4.8	2.3	3.2	3.5	1
Total (%)	100	100	100	100	100	100

These results seem logical since the ore grade of gold in Chovdar (2.39 gr/tonne) can be characterized as high for an open-pit mining operation and the average grade for an underground mining operation. The criticality of gold is prioritized to be high enough; its value as a metal makes it always a critical commodity of great strategic importance. It is interesting to note that the technological parameters are low. This can be explained by the fact that the recovery rate is already high enough, and the metallurgical tests and evaluations for the extended processing plant show excellent results.

Data and information derived from the evaluation of gold were also considered for the assessment of silver. This kind of information includes the facts that gold will most probably remain the main product at Chovdar, all costs will be covered by gold production, and silver will continue being shipped together with gold for refinement to Switzerland. Like gold, silver's price and ore grade are essential parameters to its criticality. The latter is higher than that of gold because of silver's by-production dependence on gold. Nevertheless, the grade is high enough to make silver production efficient and is combined with the commodity's importance. Silver may not be as powerful as gold, but it is also considered a strategic metal. The existence of toxic compounds during processing is also of notable priority.

Copper was the next metal to be evaluated as a potential product at Chovdar, considering the evaluation results of both gold and silver. Importance is given to the increased concentration of copper in the sulfide phase of the deposit and the fact that there is a high copper zone present in this phase.

In the oxidized phase of production, copper has been characterized as waste, rather than as a product. Hence, the economic parameters seem to be the most important, and the ore grade of the commodity is the most significant parameter by far. The price of copper will also play a role in the evaluation, whereas the extraction costs are mainly covered by the main product (gold) and are of less importance. The technological factors, and particularly the recovery rate, also have a significant weight. This makes sense since the higher the recovery of copper, the greater its chances of creating profit for the company.

Judging by the weights attributed to the parameters, it is evident that copper will be treated differently than gold and silver. Copper has low criticality and high availability as a metal worldwide, and its economic balance is the determining factor when deciding its production. The increase in concentration cannot go unnoticed, and is highlighted in the prioritization of the parameters.

The next metal, the concentration of which is increasing in the sulfide phase of the deposit, is iron. In this case, the ore grade elevation may not be as high as it is for copper, and there no high iron zone is identified. Nonetheless, the concentration is also high enough to attract interest and proceed with evaluating this commodity. The same procedure is followed for assessing iron, considering the boundary conditions at Chovdar, the market prices for iron, its importance and availability as a metal, and the potential environmental concerns that its production might raise.

Similarly, the most critical parameters for iron, as for copper, are the economic criteria, followed by the market criteria. The ore grade is the most significant factor, and the price of iron is ranked second. However, the third most crucial parameter is the impact that iron production is expected to have in the local markets.

This result is due to the wide variety and diversity of applications that iron has in daily products and services in local societies. The metallurgical process of iron is well known and can be applied near a mine site; thus, the produced iron could be channeled to the local markets, thus reducing the logistic costs. Nevertheless, the price and ore grade of iron combined with the additional extraction costs will be the main determining factors for its classification as a by-product or waste in this project.

Aluminum was assessed next. The resemblance to the properties of iron both as a commodity in general and as a potential product at Chovdar is remarkable, and so are the evaluation results. The increase in concentration for aluminum seems to be greater than that for iron, yet not significantly different.

Once again, the economic parameters seem to play a significant role when deciding whether to produce aluminum. The market conditions follow in percentage terms, and the remaining three categories (technological, environmental, and socio-political parameters) are of equally lower importance in this case. Following the same pattern, the essential parameters are the ore grade of aluminum and its price in global markets. The locality is also evaluated as a crucial parameter, followed by the additional extraction costs and the recovery rate of aluminum.

Generally, aluminum has a much higher price as a commodity than iron. In addition, the ore grade of aluminum at Chovdar is also higher than that of iron. Consequently, even though the evaluation parameters have the same weights, the evaluation of the options with respect to the parameters led to slightly different preference results.

Mercury was the last of the commodities to be evaluated in this case study using the multi-criteria decision tool. Unlike the previous metals discussed, mercury has a different treatment and production evaluation. The same group of parameters is implemented in the tool, to be evaluated concerning the properties of mercury in the Chovdar mining project, in addition to the general conditions that govern the treatment of this metal globally.

Contrary to the evaluation results in the previous paragraphs, the most important parameters, in this case, are the environmental parameters, followed by the sociopolitical parameters. The technological factors have an observable percentage. More specifically, the most significant parameters overall are the presence of toxic compounds, the production of waste, the legislation status that governs the production and treatment of mercury and, of course, the social acceptance of having it as a product or treating it as a waste.

These results are different from those discussed above regarding the other commodities. For example, the price of mercury and its marketability are not as important. The recovery rate is an essential factor, but not in terms of yielding more profit. In this scenario, the higher the recovery of mercury from the ore and tailings, the less the risk of environmental contamination. As already discussed in this work, mercury must be produced as a by-product to preserve the surrounding ecosystem, follow the rules, and meet the social requirements. In addition, when extracted and shipped off-site, the costs needed to treat mercury as waste in the tailings are eliminated, and thus can be considered an indirect profit.

6.3. Comparative Analysis of the Results

Overall, six commodities were evaluated individually but under the same circumstances and considering the same conditions of the Chovdar project. The results are rational and detailed enough, given the difficulties of deriving data when no actual economic assessments have been conducted to date for the second phase of exploitation at Chovdar.

Gold remains the leading product of the mine (Table 9), since no other commodity is classified as a co-product or will cover all the extraction costs. The transition to underground mining operations will increase the operating costs; however, gold production is expected to yield a significant profit for the company. Its strategic importance is essential not only for the relatively remote area, but also for the state of Azerbaijan. Hence, the mining project enjoys the government's trust and the society's acceptance.

Table 9. Comparative analysis of production results.

Commodities	Primary Product	Co-Product	By-Product	Waste
Gold	79.6%	-	15.5%	4.9%
Silver	-	34.4%	61.6%	4.0%
Copper	-	12.6%	49.1%	38.3%
Iron	-	13.5%	40.6%	45.9%
Aluminum	-	13.9%	43.5%	42.6%
Mercury	-	8.4%	51.9%	39.7%

Silver will be produced again as a by-product and significantly contribute to the project's revenues. Most likely, copper will be the second by-product after silver. Although the by-product option has the highest percentage, it does not have an absolute majority, indicating that a more detailed and careful evaluation must be made to decide whether copper can be feasibly extracted. Contrary to the results for copper, iron is preferably classified as waste. Nevertheless, a more detailed economic analysis is also needed for this commodity and market analysis should be undertaken to investigate the product sales prospects in the local markets. Aluminum is also not classified as a by-product or waste, although there is a slightly higher preference for it being produced as a by-product than iron. Accordingly, detailed economic and market analyses also need to be conducted for this potential product. Finally, mercury is treated differently, and the respective results justify its classification as a by-product, with a far more preferable 52%.

This last result justifies the scope for developing this production decision tool and the attempt to determine all the potential parameters, in addition to the economic parameters, that can affect the production of a commodity. The percentages of 4.9% in the preferences for gold as waste or 8.4% for mercury as a co-product are worth mentioning. These can be attributed to two reasons: the first is the using of all parameters, even the less relevant ones, in this evaluation. In these assessments, the options of co-product, by-product, or waste were equally important. Although these parameters have very low weights, their overall sum yields a higher-than-expected percentage for the option. The second reason is the lack of accurate data and information that would allow decision makers to make more precise cross-comparisons.

Compared to the Excel workbook outcomes, the results derived from the computational tool were very similar, if not identical. The insignificant differences may be attributed to the number of decimal places applied in the calculations. Nevertheless, the similarity of results justifies the efficiency of the computational tool. Python has gained traction over recent years and the quote that "Python is the new Excel" is becoming more frequent.

7. Conclusions

This work managed a large amount of information and data sets to identify and classify 18 parameters that can impact the determination of co- and by-products in a mining project. This list is not exhaustive; criteria can be added or removed, given the conditions governing each project under examination.

Using Python and a GUI, an evaluation tool was developed based on a multi-criteria-decision analysis model to assess the production perspectives in polymetallic projects. The tool allows for fast and efficient calculations, and the variation in the parameters is not an impediment. An advantage of the developed tool is that it considers more diverse parameters and yields detailed results, not only for the final options, but also for the importance of all parameters and those having the highest impact when evaluating each commodity.

To reduce subjectivity in decision making, a careful assessment needs to be made each time the tool is used concerning the boundary conditions of each project and the precision of the data and information provided for the evaluations.

The tool's efficiency was tested by implementing data from a polymetallic deposit in Chovdar, Azerbaijan. In this project, operations are transitioning from surface to underground, in which the mineralogy is also changing. Hence, a re-evaluation of the perspectives of the included metals aiming at their production feasibility was deemed necessary.

Overall, the evaluation results from the tool justified the production of gold, silver, and mercury that is already taking place in Chovdar, indicating that the tool works efficiently and can be used accordingly for the other commodities. Therefore, the results for the remaining potential products indicate the approach for the company to investigate whether any of these metals can become by-products of the project.

Mining companies, industry consultants, academics, and other stakeholders could use the developed tool in the assessments of several polymetallic projects. In this manner, the tool can be further tested and optimized, and the use of additional parameters can be determined. Consequently, the necessity of producing minor metals will be further highlighted, not only with words but with the demonstration of detailed results and percentages.

Author Contributions: Conceptualization, A.M. and G.B.; methodology, A.M., G.B. and M.A.I.; software, M.A.I.; validation, A.M., G.B. and H.M.; formal analysis, A.M., G.B., M.A.I., H.M. and M.H.; investigation, A.M. and G.B.; data curation, A.M. and G.B.; writing—original draft preparation, A.M. and G.B.; writing—review and editing, H.M. and M.H.; visualization, G.B. and M.A.I.; supervision, H.M. and M.H. All authors have read and agreed to the published version of the manuscript.

Funding: This research received no external funding.

Conflicts of Interest: The authors declare no conflict of interest.

References

1. OECD. *Global Material Resources Outlook to 2060: Economic Drivers and Environmental Consequences*; OECD Publishing: Paris, France, 2019. [CrossRef]
2. Nassar, N.T.; Graedel, T.E.; Harper, E.M. By-product metals are technologically essential but have problematic supply. *Sci. Adv.* **2015**, *1*, e1400180. [CrossRef] [PubMed]
3. Mookherjee, A.; Panigrahi, M.K. Reserve base in relation to crustal abundance of metals: Another look. *J. Geochem. Explor.* **1994**, *51*, 1–9. [CrossRef]
4. Skinner, B.J. Earth resources. *Proc. Natl. Acad. Sci. USA* **1979**, *76*, 4212–4217. [CrossRef] [PubMed]
5. Campbell, G.A. The role of co-products in stabilizing the metal mining industry. *Resour. Policy* **1985**, *11*, 267–274. [CrossRef]
6. Wellmer, F.-W.; Hannak, W.; Krauss, U.; Thormann, A. Deposits of rare metals. In Proceedings of the 5th International Symposium, Hannover, Germany, 19–21 October 1988; pp. 71–122.
7. Verhoef, E.V.; Dijkema, G.P.J.; Reuter, M.A. Process, Knowledge, System Dynamics, and Metal Ecology. *J. Ind. Ecol.* **2004**, *8*, 23–43. [CrossRef]
8. Reuter, M.A.; Heiskanen, K.; Boin, U.M.J.; Schaik, A.; Verhoef, E.V.V.; Yang, Y.; Georgialli, G. *The Metrics of Material and Metal Ecology*; Elsevier: Amsterdam, The Netherlands, 2005; p. 706.
9. Buchert, M.; Schüler, D.; Bleher, D. *Critical Metals for Future Sustainable Technologies and Their Recycling Potential*; Öko-Institut e.V. (UNEP): Freiburg, Germany, 2009.
10. Willis, P.; Chapman, A.; Fryer, A. Study of By-Products of Copper, Lead, Zinc and Nickel, Oakdene Hollins. 2012. Available online: www.oakdenehollins.com/s/By-Products_Report_Executive_Summary_version_on_website.pdf (accessed on 10 March 2020).
11. Wellmer, F.-W.; Hagelüken, C. The Feedback Control Cycle of Mineral Supply, Increase of Raw Material Efficiency, and Sustainable Development. *Minerals* **2015**, *5*, 815–836. [CrossRef]
12. Mudd, G.M.; Wenga, Z.; Jowittb, S.M.; Turnbullc, I.D.; Graedelc, T.E. Quantifying the recoverable resources of by-product metals: The case of cobalt. *Int. J. Serv. Sci.* **2013**, *55*, 87–98. [CrossRef]

13. Frenzel, M.; Tolosana-Delgado, R.; Gutzmer, J. Assessing the supply potential of high-tech metals—A general method. *Resour. Policy* **2015**, *46*, 45–58. [CrossRef]
14. Frenzel, M.; Mikolajczak, C.; Gutzmer, J.; Reuter, M.A. Quantifying the relative availability of high-tech by-product metals—The cases of gallium, germanium and indium. *Resour. Policy* **2017**, *52*, 327–335. [CrossRef]
15. Schaik, A.V.; Reuter, M.A. The optimization of end-of-life vehicle recycling in the European Union. *JOM* **2004**, *56*, 39–43. [CrossRef]
16. Binnemans, K.; Jones, P.T.; Blanpain, B.; Gerven, T.V.; Pontikes, Y. Towards zero-waste valorisation of rare-earth-containing industrial process residues: A critical review. *J. Clean. Prod.* **2015**, *99*, 17–38. [CrossRef]
17. Peelman, S.; Kooijam, D.; Sietsma, J.; Yang, Y. Hydrometallurgical recovery of rare earth elements from mine tailings and WEEE. *J. Sustain. Metall.* **2018**, *4*, 367–377. [CrossRef]
18. Alvarado-Hernández, L.; Lapidus, G.T.; González, F. Recovery of rare earths from waste cathode ray tube (CRT) phosphor powder with organic and inorganic ligands. *Waste Manag.* **2019**, *95*, 53–58. [CrossRef] [PubMed]
19. Valero, A.; Dominguez, A.; Valero, A. Exergy cost allocation of by-products in the mining and metallurgical industry. *Resour. Conserv. Recycl.* **2015**, *102*, 128–142. [CrossRef]
20. Renner, S.; Wellmer, F.W. Volatility drivers on the metal market and exposure of producing countries. *Miner. Econ.* **2019**, *33*, 311–340. [CrossRef]
21. Hitch, M.; Ballantyne, S.M.; Hindle, S.R. Revaluing mine waste rock for carbon capture and storage. *Int. J. Min. Reclam. Environ.* **2009**, *24*, 64–79. [CrossRef]
22. Tilton, J.E.; Guzmān, J.I. *Mineral Economics and Policy*, 1st ed.; RFF Press: New York, NY, USA, 2016; ISBN 978-1-138-83895-6.
23. Redlinger, M.; Eggert, R. Volatility of by-product metal and mineral prices. *Resour. Policy* **2016**, *47*, 69–77. [CrossRef]
24. Deselnicu, D.C.; Militaru, G.; Deselnicu, V.; Zainescu, G.; Albu, L. Towards a circular economy—A zero waste programme for Europe. In Proceedings of the 7th International Conference on Advanced Materials and Systems, Bucharest, Romania, 18–20 October 2018.
25. Vilella, M. Zero Waste Circular Economy—A Systemic Game-Changer to Climate Change. Heinrich Böll Stiftung. Publication Series Ecology. V 44.3. 2018. Available online: www.boell.de/sites/default/files/radical_realism_for_climate_justice_volume_44_3.pdf (accessed on 3 April 2020).
26. Saaty, T.L. *The Analytic Hierarchy Process: Planning, Priority Setting, Resource Allocation*, 1st ed.; McGraw Hill International: New York, NY, USA, 1980; p. 287.
27. Barakos, G.; Mischo, H.; Gutzmer, J. Legislation; Challenges and Policy Strategies: In Search for a Regulatory Framework for Sustainable Development in the Rare Earths Mining Industry. In Proceedings of the SME 2016, Phoenix, AZ, USA, 21–24 February 2016.
28. Patterson, S.H.; Kurtz, H.F.; Olson, J.C.; Neeley, C.L. *World Bauxite Resources*, 1st ed.; U.S. Geological Survey: Washington, DC, USA, 1986.
29. Rivera, R.M.; Ulenaers, B.; Ounoughene, G.; Binnemans, K.; Gerven, T.V. Extraction of rare earths from bauxite residue (red mud) by dry digestion followed by water leaching. *Miner. Eng.* **2018**, *119*, 82–92. [CrossRef]
30. Akcil, A.; Akhmadiyeva, N.; Abdulvaliyev, R.; Meshram, A.; Meshram, P. Overview on extraction and separation of rare earth elements from red mud: Focus on scandium overview on extraction and separation of rare earth elements from red mud: Focus on scandium. *Miner. Process. Extr. Metall. Rev.* **2017**, *39*, 145–151. [CrossRef]
31. Borra, C.R.; Pontikes, Y.; Binnemans, K.; Gerven, T.V. Leaching of rare earths from bauxite residue (red mud). *Miner. Eng.* **2015**, *76*, 20–27. [CrossRef]
32. Vind, J.; Malfliet, A.; Bonomi, C.; Paiste, P.; Sajo, I.E.; Blanpain, B.; Tkaczyk, A.H.; Vassiliadou, V.; Panias, D. Modes of occurrences of scandium in Greek bauxite and bauxite residue. *Miner. Eng.* **2018**, *123*, 35–48. [CrossRef]
33. Zhang, N.; Li, H.; Liu, X. Recovery of scandium from bauxite residue—Red mud: A review. *Rare Met.* **2016**, *35*, 887–900. [CrossRef]
34. Olivetti, E.A.; Ceder, G.; Gaustad, G.G.; Fu, X. Lithium-Ion Battery Supply Chain Considerations: Analysis of Potential Bottlenecks in Critical Metals. *Joule* **2017**, *1*, 225–228. [CrossRef]
35. Battery Raw Materials, Commodity Profiles, British Geological Survey. 2018. Available online: www.bgs.ac.uk/downloads/start.cfm?id=3403 (accessed on 27 February 2020).
36. Gounaris, K. Applicability of Digital Tools for the Assessment of the Global Mining Industry of Battery Raw Materials. Master's Thesis, TU Bergakademie Freiberg, Freiberg, Germany, 2019.
37. Rio Tinto: The Boron Mining Project. Available online: www.riotinto.com/en/operations/us/boron (accessed on 15 April 2020).
38. López, F.A.; Garcia Diaz, I.; Rodrigez Largo, O.; Polonio, F.; Llorens, T. Recovery and Purification of Tin from Tailings from the Penouta Sn–Ta–Nb Deposit. *Minerals* **2018**, *8*, 20. [CrossRef]
39. Nadeif, A.; Taha, Y.; Bouzahzah, H.; Hakkou, R.; Benzaazoua, M. Desulfurization of the old tailings at the Au-Ag-Cu Tiouit Mine (Anti-Atlas Morocco). *Minerals* **2019**, *9*, 401. [CrossRef]
40. Garcia-Guinea, J.; Harffy, M. Mercury mining: Profit or loss? *Nature* **1997**, *390*, 112–113. [CrossRef]
41. Gill, G.A.; Bruland, K.W. Mercury speciation in surface freshwater systems in California and other areas. *Environ. Sci. Technol.* **1990**, *24*, 1392–1400. [CrossRef]
42. Malm, O.; Pfeiffer, W.C.; Souza, C.M.M.; Reuther, R. Mercury pollution due to gold mining in the Madeira River basin, Brazil. *AMBIO* **1990**, *19*, 11–15.

43. Gomez, M.G.; Klink, J.D.C.; Boffetta, P.; Espanol, S.; Sällsten, G.; Quintana, J.G. Exposure to mercury in the mine of Almaden. *Occup. Environ. Med.* **2007**, *64*, 389–395. [CrossRef]
44. Pirrone, N.; Cinnirela, S.; Feng, X.; Finkelman, R.B.; Friedli, H.R.; Leaner, J.; Mason, R.; Mukherjee, A.B.; Stracher, G.; Streets, D.G.; et al. Global mercury emissions to the atmosphere from natural and anthropogenic sources. In *Mercury Fate and Transport in the Global Atmosphere: Emissions, Measurements and Models*, 1st ed.; Pirrone, N., Mason, R., Eds.; Springer: New York, NY, USA, 2009; pp. 3–50. [CrossRef]
45. Dionyssomarble: Dionyssos—Pentelicon Commercial & Industrial Marble. Available online: www.dionyssomarble.com/en/ (accessed on 18 February 2020).
46. Baar, C.A. *Applied Salt-Rock Mechanics: The In-Situ Behavior of Salt Rocks*, 1st ed.; Elsevier: Amsterdam, The Netherlands, 1977; p. 293.
47. Blackburn, R.R.; Grane, E.J.M.; Chappelow, C.C.; Harwood, D.W. *Development of Anti-Icing Technology*; Strategic Highway Research Program; National Research Council: Washington, DC, USA, 1994. Available online: http://onlinepubs.trb.org/onlinepubs/shrp/SHRP-H-385.pdf (accessed on 25 April 2020).
48. Komulainen, V. Road maintenance in Finland and Germany. Bachelor's Thesis, Saimaa University of Applied Sciences, Lappeenranta, Finland, 2011.
49. GSES: From the 'Glückauf' Potash Mine to a Modern Service Provider in Mining-Related Field. Available online: www.gses.de/history.html (accessed on 13 March 2020).
50. K+S: After 120 Years, Potash Mining Ended: The History of the Sigmundshall Plant. Available online: www.kpluss.com/en-us/about-ks/locations/europe/sigmundshall/ (accessed on 13 March 2020).
51. K+S: One of the World's Largest and Most Modern Potash Mines Near Magdeburg: The History of the Zielitz Potash Plant. Available online: www.kpluss.com/en-us/about-ks/locations/europe/zielitz/ (accessed on 13 March 2020).
52. Mischo, H. *Entdecker unter Tage 1919–2019: 100 Jahre Forschungs- und Lehrbergwerk*, 1st ed.; Mitteldeutscher Verlag: Halle, Germany, 2019; ISBN 978-3963111907. (In German)
53. Bayer, M. *Die Himmelfahrt Fundgrube: Ein Führer Durch das Lehr- und Besucherbergwerk der TU Bergakademie Freiberg*, 1st ed.; TU Bergakademie Freiberg: Freiberg, Gemany, 1997. (In German)
54. Grosse Kreisstadt Freital: Bergbau & Industrie: Freital—Gegründet auf Kohle und Stahl. Available online: www.freital.de/Unsere-Stadt/Stadtportr%C3%A4t-Stadtgeschichte/Stadtchronik/Bergbau-Industrie (accessed on 1 April 2020). (In German).
55. Graedel, T.E.; Harper, E.M.; Nassar, N.; Nuss, P.; Reck, B.K. Criticality of metals and metalloids. *Proc. Natl. Acad. Sci. USA* **2015**, *112*, 4257–4262. [CrossRef] [PubMed]
56. Lööw, J.; Abrahamsson, L.; Johansson, J. Mining 4.0—The Impact of New Technology from a Work Place Perspective. *Min. Metall. Explor.* **2019**, *36*, 701–707. [CrossRef]
57. Tchounwou, P.B.; Yedjou, C.G.; Patlolla, A.K.; Sutton, D.J. Heavy metals toxicity and the environment. *Mol. Clin. Environ. Toxicol.* **2012**, *101*, 133–164. [CrossRef]
58. Beamish, B.B.; Vance, W.E. Greenhouse gas contributions from coal mining in Australia and New Zealand. *J. R. Soc. N. Z.* **1992**, *22*, 153–156. [CrossRef]
59. Pandey, B.; Gautam, M.; Agrawal, M. Chapter 10—Greenhouse Gas Emissions From Coal Mining Activities and Their Possible Mitigation Strategies. *Environ. Carbon Footpr.* **2018**, 259–294. [CrossRef]
60. Borden, R.K. Waste disposal and contamination management. In *SME Mining Engineering Handbook*, 3rd ed.; Darling, P., Ed.; Society for Mining, Metallurgy, and Exploration: Englewood, NJ, USA, 2011; pp. 1733–1751.
61. Barakos, G.; Mischo, H. Insertion of the social license to operate into the early evaluation of technical and economic aspects of mining projects: Experiences from the Norra Kärr and Bokan Dotson rare earth element projects. *Extr. Ind. Soc.* **2021**, *8*, 100814. [CrossRef]
62. Lesser, P.; Gugerell, K.; Poelzer, G.; Hitch, M.; Tost, M. European mining and the social license to operate. *Extr. Ind. Soc.* **2021**, *8*, 100787. [CrossRef]
63. Miller, G.A. The magical number seven, plus or minus two: Some limits on our capacity for processing information. *Psychol. Rev.* **1956**, *63*, 81–97. [CrossRef] [PubMed]
64. Saaty, T.L.; Ozdemir, M.S. Why the magic number seven plus or minus two. *Math. Comput. Model.* **2003**, *38*, 233–244. [CrossRef]
65. Ishizaka, A.; Nemery, P. *Multi-Criteria Decision Analysis*; John Willey & Sons: Hoboken, NJ, USA, 2013.
66. Chen, Y. Multi Criteria Decision Analysis: Classification Problems and Solutions. Ph.D. Thesis, University of Waterloo, Waterloo, ON, Canada, 2006.
67. Coyle, G. The analytic hierarchy process (AHP). In *Practical Strategy: Structured Tools and Techniques*; Coyle, G., Ed.; Pearson Education: London, UK, 2004.
68. Saaty, T.L. Decision making with the analytic hierarchy process. *Int. J. Serv. Sci.* **2008**, *1*, 83–98. [CrossRef]
69. Kappes, Cassiday & Associates. *Chovdar Gold Deposit, Khalnar Administrative District. Scoping Study*; Technical Report for the Chovdar gold project; Kappes, Cassiday & Associates: Reno, NV, USA, 2011.
70. Abdullayeva, S.F.; Baba-Zadeh, V.M.; Musayev, S.D.; Akhmedov, A.Z.; Imamverdiyer, N.A.; Romanko, A.E. The Material Composition and Technological Characteristics of Ores in Chovdar Gold-Ore Deposit. *Int. J. Innov. Sci. Eng. Technol.* **2016**, *3*, 132–160.
71. Chovdar Gold Mine. Available online: https://azergold.az/en (accessed on 13 March 2020).

mining

MDPI

Article

Environmental and Work Factors That Drive Fatigue of Individual Haul Truck Drivers

Elaheh Talebi [1,*], W. Pratt Rogers [1] and Frank A. Drews [2]

1 Department of Mining Engineering, University of Utah, Salt Lake City, UT 84112, USA
2 Department of Psychology, University of Utah, Salt Lake City, UT 84112, USA
* Correspondence: elaheh.talebi@utah.edu

Abstract: Many factors influence the fatigue state of human beings, and fatigue has a significant adverse effect on the health and safety of the haulage operators in the mine. Among various fatigue monitoring systems in mine operations, currently, the Percentage of Eye Closure (PERCLOS) is common. However, work and other environmental factors influence the fatigue state of haul truck drivers; PERCLOS systems do not consider these factors in their modeling of fatigue. Therefore, modeling work and environmental factors' impact on individual operations fatigue state could yield interesting insights into managing fatigue. This study provides an approach of using operational data sets to find the leading indicators of the operators' fatigue. A machine learning algorithm is used to model the fatigue of the individual. eXtreme Gradient Boosting (XGBoost) algorithm is chosen for this model because of its efficiency, accuracy, and feasibility, which integrates multiple tree models and has stronger interpretability. A significant number of negative and positive samples are created from the available data to increase the number of datasets. Then, the results are compared with other existing models. A selected algorithm, along with a big data set was able to create a comprehensive model. The model was able to find the importance of the individual factors along with work and environmental factors among operational data sets.

Keywords: machine learning; XGBoost algorithm; PERCLOS system; fatigue

Citation: Talebi, E.; Rogers, W.P.; Drews, F.A. Environmental and Work Factors That Drive Fatigue of Individual Haul Truck Drivers. *Mining* **2022**, 2, 542–565. https://doi.org/10.3390/mining2030029

Academic Editors: Juan M. Menéndez-Aguado, Oscar Jaime Restrepo Baena and Jessica M. Smith

Received: 22 July 2022
Accepted: 23 August 2022
Published: 26 August 2022

1. Introduction

Fatigue is an occupational hazard and can be attributed to the health and safety of the worker. It affects the health and safety of both the employees and their colleagues adversely. Fatigue is a complex phenomenon that can be associated with many factors. Fatigue can be defined as a state of feeling tired, weary, or sleepy that results from prolonged physical or mental work, extended periods of anxiety, exposure to harsh environments, or lack of sleep [1]. Fatigue varies from weakened function of alertness during tasks to drowsiness, micro-sleep or completely falling asleep. It can affect worker performance and impair their mental alertness, which can cause dangerous errors [1]. Fatigue presents several challenges for the mining industry. Various accidents have been reported at mine operations, which could be associated with the loss of control due to the fatigue and sleepiness of mineworkers [2]. The mining industry is certainly not alone in facing the challenge of addressing worker fatigue. In fact, many of the characteristics of fatigue in the mining industry mirror the similarities of fatigue in other industries. Hence, any fatigue management applications, training, or interventions from other industries can be borrowed and applied to mining. However, some have argued that mining, in particular, is especially susceptible to increases in the presence of fatigue due to the multifaceted combination of factors in mining environments associated with fatigue: dim lighting; limited visual acuity; hot temperatures; loud noise; highly repetitive, sustained, and monotonous tasks; shiftwork; long work hours; long commute times due to mine site remoteness; early morning awakenings; and generally poor sleep habits [3]. Although the full burden of fatigue on mineworkers has not yet been measured, some technologies can monitor operator fatigue, such as the video-based technology, percentage of eyelid closure over the pupil over time (PERCLOS) system, and Electroencephalography (EEG) cap. While such technologies can detect the fatigue of the operator, they do not necessarily prevent or mitigate fatigue from happening [3]. These technologies have their pros and cons. Video-based technology (PERCLOS) uses eyelid closure as the main measure to detect driver fatigue. However, it has some drawbacks that limit its success, such as darkness limitations and practical hurdles like the distraction of the drivers [4]. On the other hand, the EEG cap uses electrodes to get the signal from the brain and translate it to the status of the driver fatigue. These electrodes and caps have been shown to cause discomfort for drivers, limiting their widespread adoption [5]. Therefore, they are not suitable for long-term monitoring. There is another technology called WOMBATT (Worldwide Online Monitoring By Alerting Tired Travelers), can determine a person's level of fatigue by analyzing a recording voice [6]. PERCLOS, EEG, WOMBATT and other technologies can detect fatigue, but they cannot recognize factors that affect fatigue such as work-demand factors.

The Job Demand-Resources model (JDR) of worker stress and health shows that too much job demand and not enough resources can result in injuries due to physiological and physical costs [3]. The core part of the JDR model is related to risk factors associated with job stress. These factors can be classified into two general categories, job demands and job resources [7–9]. Job demands refer to the physical, psychological, social, or organizational aspects of the job that require physical and psychological effort or skills. However, job demands are not necessarily negative; they may turn into job stressors when those demands require high effort from the employee, which is hard to recover [7]. Job resources refer to the physical, psychological, social, or organizational aspects of the job that are functional in achieving work goals, decreasing job demands, or stimulating personal growth, learning, and development [7–9].

2. Previous Studies

Fitness for duty in mining is an important issue which is affected by individual's physical and psychological fitness. Fatigue is one of the driver of fitness for duty in mining, which greatly is caused by excessive work hours and shiftwork [10,11]. Fatigue in the workplace often results in a reduction in worker performance. Fatigue must be controlled and managed since it causes significant short-term and long-term risks [12–15]. Other than the health and safety consequences on workers, fatigue can result in damage or loss of valuable mine equipment like haul trucks. So, the mining industry measures operational risk losses to estimate capital allocation and manage operational risks [16,17].

Drews et al. (2020) studied fatigue in the mining industry and mentioned that fatigue in the mining industry is different from other industries because of the specific environmental factors in the mining industry [18]. They also provided some other factors that drive fatigue like repetitive and monotonous tasks, long work hours, shiftwork, sleep deprivation, dim lighting, limited visual acuity, hot temperatures, and loud noise [18]. Multiple psychological and physiological issues impacted the fatigue of the workers, which makes fatigue management difficult. Some technologies can monitor drivers' fatigue, such as tracking eye movement and head orientation (PERCLOS) or hard hats with electroencephalogram (EEG) activity tracking, with their pros and cons. However, considering these technologies, other studies show that there is no obvious approach to control and mitigate the fatigue of workers in mine operations [3].

Machine learning (ML) can be used to predict leading indicators and help management make appropriate decisions [19,20]. ML is flexible to operate without any statistical assumptions. It also is able to identify any relationships within the phenomena and issues [19,21,22]. Previous study offers that finding leading indicators to predict fatigue in the mining industry can be useful [18]. Due to the complexity of fatigue, using machine learning (ML) algorithms on the real-time data captured from the existed technologies can be helpful to model fatigue. Such a model could identify predictive elements of workers' fatigue. Some studies are done with the collected data to predict fatigue [18]. However, a comprehensive study using a wider range of available data sets can find more possible independent variables in the model to find top predictive factors. If these factors can be used as fatigue predictive elements, they will enhance safety and health decisions in an earlier time in the fatigue cycle.

In an earlier study done by E. Talebi et al. (2021), machine learning (ML) models were created using aggregated operational data sets from a mine [23]. The findings of that study confirm that fatigue is caused by a wide variety of factors, which are very difficult to quantify. Fatigue prediction is a matter of predicting the complex interactions between human behavior and the changing work environments at mine operations [23]. The model outcome had a low R^2 value that captures relationships that quantify a relatively high amount of variance in a complex relationship. This high amount of variance is likely largely due to the difficulty of generalizing a model that can predict fatigue due to the complex psychological and physiological factors associated with fatigue at the group level. Only operational data and weather data are utilized in these models aggregated at the mine level [23].

The machine learning model selected for that analysis was a random forest (RF) regression algorithm. This algorithm was chosen because it can be applied well to a wide variety of problems with a rapid speed of training. This analytical tool shows what features of the model have higher effects on the predictions of the model and estimates how marginal changes in those features impact these predicted outcomes [23].

The model output identifies the variables that have the highest impact on all fatigue events. The previous model results offer some interesting insights into the factors that potentially cause fatigue. It shows that while it is not surprising that shift type (night or day) causes fatigue, it is interesting that maintenance processes such as unscheduled downtime and production rates, as well as other operational variables, can affect fatigue among haul truck drivers. Having identified these additional predictors for fatigue, these indicators can be used by managers to prioritize safety management efforts. However, this model was aggregated and averaged out at the mine level. Would a model at a lower level of granularity, say, at the individual level, rather than aggregated at the mine level yield better results? This was the primary difference and guiding research question for the model presented in this paper.

XGBoost algorithm is applied to data to model fatigue. This algorithm is used because it has more power to handle complicated relations. XGBoost is a powerful machine learning (ML) algorithm that has shown strong power to pick up patterns in the data and automatically tune learnable parameters. What is novel in this study compared to the previous study is a higher score of the model to predict fatigue.

3. Case Study

3.1. Data Description

This study used approximately four years of data from a single, large, operating surface mine. Table 1 shows a brief overview of the data sets that were used for modeling fatigue with details of the types of information and the range of dates. The site utilized a PERCLOS monitoring system, which used cameras to track and monitor the eye movements of haul truck drivers to model and detect fatigue. When the camera detected certain eye movements, eye closure, or blinking, the PERCLOS system can determine fatigue based on a preset model. In a situation when the eyes were closed for more than 3 s, the system alerted operators, supervisors, and dispatchers for more action. Data captured from the system was categorized based on the type of event. If the event was a micro-sleep, which was an actual fatigue event, it would be categorized as a low or critical fatigue event. Previous studies by the authors showed that fatigue events captured by fatigue monitoring systems are important indicators of fatigue [10]. Therefore, micro-sleep data was used to model fatigue for this study.

Table 1. Data sets details.

Data Source	Key Factors	Date Range
Fatigue monitoring	Micro-sleeps (low and critical fatigue)	2016–2020
Time and attendance	Hours worked, overtime, etc.	2016–2020
Fleet management system (production and status)	Production cycles, down equipment, delayed equipment, etc.	2016–2020

More details of fatigue events are shown in Table 2. This table demonstrates the number of events by type of fatigue events and the percentage of these fatigue events for comparison. The data shows more low fatigue compared to critical fatigue, representing 69% of the fatigue events that were captured by the system. All fatigue events are reviewed after recording from the fatigue monitoring system, and critical fatigue events are the ones when operators have micro-sleep, while low fatigue events are the ones that just show drowsiness.

Table 2. Count and percentage of fatigue event by type.

Fatigue Event Review Type	Number of Fatigue Events	Percentage of Fatigue
Low Fatigue	741	69%
Critical Fatigue	332	31%

Data from the fleet management system (FMS) tracked the production and status of equipment. It offers a good perspective on the job demands of haul truck drivers throughout the shift. Status event or status of the equipment can be used to determine if a piece of equipment is down for maintenance, in production activity, in standby mode, or ready for production. This information can be used to find the status of the haul truck at the time of proceeding the fatigue events. Other information in the FMS database included the load cycle data. A production cycle showed the load and dump cycles of a truck. Detailed steps were also provided, such as loading, dumping, running empty, running loaded, etc., are shown. The most important data for this study was the production cycle state of the truck when fatigue events happened.

Time and attendance data are provided to show hours worked by employees. The mine used a swipe-in/swipe-out time keeping system to process and load into a time and attendance database. A data set of attendance from this database was used to measure worked hours and overtime of the employees.

3.2. Data Pre-Processing

For the application of machine learning algorithms, data must be pre-processed in a mathematically feasible format. Therefore, data needs to be pre-processed to make it appropriate for the application of the modeling. Data pre-processing techniques included data reduction, data projection, and missing-data treatment. In data reduction, the size of the datasets decreases by means of feature selection. Data projection intends to transform all features into a conformed format and range. Missing-data treatments include deleting missing values and replacing them with the estimates if needed. Therefore, data needs to be pre-processed to make it appropriate for the application of the modeling.

3.2.1. Data Integration

Each data set was linked to the fatigue monitoring data set based on a unique key. The FMS system separates equipment status and states. These tables also had to be integrated into a complex join. Data from FMS system, including status events of the equipment, were joined to the fatigue data by using a unique key. In order to have a categorized model, positive and negative samples are created from the fatigue monitoring data set, which are from the time operators were fatigued or not. After creating samples, other data from the attendance database such as overtime, worked hours and number of cycles from the Load-Dump cycles, were attached to them. In addition, the state of the haul truck at the time of the fatigue is integrated into the samples from the Load-Dump cycles data set. Finally, positive and negative samples are integrated together for the purpose of the model.

3.2.2. Data Cleaning

All of the datasets were cleaned, and missing data removed prior to input to the model. The process of cleaning data included correcting data types, removing incorrect, duplicate, incomplete, and corrupted data. The next step is handling missing values, like replacing them with anticipated data or dropping out the whole row from the data. In some cases, unwanted data has to be dropped from the data. In addition, the type of the data sets may need to be updated. Finally, multiple datasets should be merged.

In this study, missing data either was filled out with the estimated value, or the whole rows of them were dropped. Missing values can be handled by deleting the rows having null values. The rows which are having one or more column values as null can be dropped. In the case that we want to keep the row, Random Forest algorithm is used to fill out the missing values. It looks at the same data and predicts the missing value. Data are trained by the rows that have all values and predict values for the rows we have missing values. In order to join dump and load data, load IDs were created, which were comprised of shift index, shovel ID, Truck ID and arrive time. Some of these arrive times were missing; in which case an estimate was used based on the load-cycle data at the closest date and time. In the loads-dumps data set, almost 500,000 records were missing. Moreover, some of the data types and formats were changed for modeling purposes. Additionally, unwanted, duplicated, corrupted, and incorrect data were omitted from the data source.

3.2.3. Negative and Positive Samples

In this section, the process of how samples are created is explained. In this study, the categorical machine learning (ML) model is used. It means that data includes different data categories, which are positive and negative samples. The model predicts if the data is related to the positive sample, which means fatigue happened, or negative samples, which are related to the time without fatigue. Positive samples are derived from the fatigue monitoring data sets when a fatigue event is flagged by the system.

Negative samples were made from time frame when fatigue events did not happen. This sampling was done for each employee and equipment. In the process of data engineering, fatigue data is merged with status event data. Two factors are used for making negative samples. First, in order to find the number of samples, we looked at the ratio of the time frame that fatigue did not happen during a shift time for each employee. Second, we looked at each status event in a shift in a way that we have at least one sample for each status event. Therefore, these negative samples are created in an acceptable proportion ratio for the time frame that fatigue did not happen in a shift time and for each status event.

After these samples are created, other variables like time and attendance data, number of cycles, and overtime are merged with these samples. Moreover, other feature engineering is done for these samples. Finally, negative and positive samples are combined to have a big data set for the purpose of making the model. More details of what is done on the data in this process are explained in the feature engineering section.

3.2.4. Feature Engineering

Features are the numerical or categorical variables from the data sets that can determine the model prediction. They are independent, and ideally, there is little to no correlation between the features. Feature engineering has a vital role in data analysis and machine learning. Feature engineering meets a need for the generation and selection of useful features. It includes different steps of engineering as will be explained. Feature transformation, feature generation, and feature extraction are about making a feature from existing features [24]. Feature selection is about selecting a small set of features from the datasets to make it computationally feasible to use in a certain algorithm. Feature analysis and evaluation are the processes of evaluating the usefulness of the features, which is usually a part of feature selection [24].

In this study, all datasets are engineered in an appropriate way to be used in an XGBoost algorithm [25]. Fatigue data provided from the fatigue monitoring system were reviewed and divided into different categories. Among them, micro-sleeps and drowsiness were identified as the fatigue events of workers with low and critical fatigue levels. They were dependent variables of the model. All other available data like fleet management data, production cycles, and time and attendance were modeled as predictors and independent variables of the model. All features and variables used in different iterations of the modeling are shown in Table 3.

Table 3. List of the variables based on the data source.

Data Source	Variables	Data Type and Example Data
Time and attendance	Employee ID	Integer (5 to 89,021)
	Common Equipment ID	Integer (205 to 841)
	Year	Integer (2014 to 2020)
	Month	Integer (1 to 12)
	Week	Integer (1 to 54)
	Day	Integer (1 to 31)
	Day of week	Integer (1 to 7)
	Day of year	Integer (1 to 366)
	Shift is end of month	Categorical Integer (0 and 1)
	Shift is start of month	Categorical Integer (0 and 1)
	Shift is end of quarter	Categorical Integer (0 and 1)
	Shift is start of quarter	Categorical Integer (0 and 1)
	Shift is end of year	Categorical Integer (0 and 1)
	Shift is start of year	Categorical Integer (0 and 1)
	Worked hour	Integer (0 to 13)
	Overtime	Integer (0 to 1)
	Overtime of previous shift	Integer (0 to 1)
Fleet management system (production and status)	Status—Delay	Categorical Integer (0 and 1)
	Status—Down	Categorical Integer (0 and 1)
	Status—Ready	Categorical Integer (0 and 1)
	Status—Standby	Categorical Integer (0 and 1)
	Number of cycles	Integer (1 to 121)
	Number of cycles of previous shift	Integer (1 to 121)
	Work Type Scheduled	Categorical Integer (0 and 1)
	Work type Unscheduled	Categorical Integer (0 and 1)
	Fleet_HTE—KOM960	Categorical Integer (0 and 1)
	Fleet_HTE—LEB282	Categorical Integer (0 and 1)
	Fleet_HTM—CAT793	Categorical Integer (0 and 1)
	Reason Description (Included of 70 different variables)	Categorical Integer (0 and 1)
	State of Truck—Empty Driving	Categorical Integer (0 and 1)
	State of Truck—Full Driving	Categorical Integer (0 and 1)
	State of Truck—Loading	Categorical Integer (0 and 1)
	State of Truck—None	Categorical Integer (0 and 1)
	State of Truck—In Queue Waiting for Loading	Categorical Integer (0 and 1)
	State of Truck—Spotting for Load	Categorical Integer (0 and 1)
	Status Event Duration	Integer (2 to 167,932)
Fatigue monitoring system	Is Fatigue	Categorical Integer (0 to 1)

3.2.5. Features

This model was aggregated at the individual level (haul truck operators), with positive and negative samples as it is explained before. Since this model is a categorical model, a dependent variable is added to the data to show if a row of data is related to the fatigue event or not (True or False). Therefore, data are categorized into positive and negative samples. Positive samples are the rows of input data of the model that show actual fatigue events. Negative samples are the rows of data that are not fatigue events.

All the features were created for both positive and negative samples. Two different features were provided from the time and attendance data sets, overtime of current or previous shift and worked hours of the same shift for each sample. Moreover, some variables like day, week, month, and year were built from the attendance datasets. Loads and dumps data were part of the fleet management data represented by the load-dump cycle of the haul truck. They were used to create a feature of the number of cycles for the current shift and previous shift. Another feature that was provided from the cycle data is the states of the haul truck for each positive and negative sample. They were categorized into Queue Waiting for Load, Spotting for Load, Loading, Full Driving, Dumping, and Empty Driving. Status events datasets are used to find the status of the haul truck for the samples, such as Unscheduled work, Scheduled work, Ready, Delay, Down, Standby, and Event duration.

3.3. Data Visualization

Before creating any machine learning model and after data cleaning, it is necessary to do exploratory analysis. First, data should be examined before training the model. The Pandas library in Python was used to load data into a DataFrame structure for further manipulation. Then, some basic statistical analyses were generated, for example, the distribution of each countable variable (Figure 1). Other analyses produced linear correlations to observe the relationship between independent variables. Figure 2 displays the significant correlation of the variables.

As Figure 1 shows, 75% of the employees have 0 5 to 1.5 h of overtime, which can be seen from the worked hour data as well. On the other hand, the average number of cycles from the previous shift is almost 20 for most of the data. It also shows the right-skewed distribution, which identifies that most of the data have more than 50 cycles of the previous shift. Same data as overtime data from the worked hour graph shows that more than 75% of employees worked more than 12 h a shift Other graphs display which day, month, and year have the higher rate of fatigue. The last graph shows which equipment ID has a higher rate of fatigue compared to others. Figure 2 demonstrates that some of the independent variables have a positive or negative correlation with more than $R^2 = 0.7$, which has darker red or blue color. Therefore, one of them is removed from the model to reduce the possibility of overfitting.

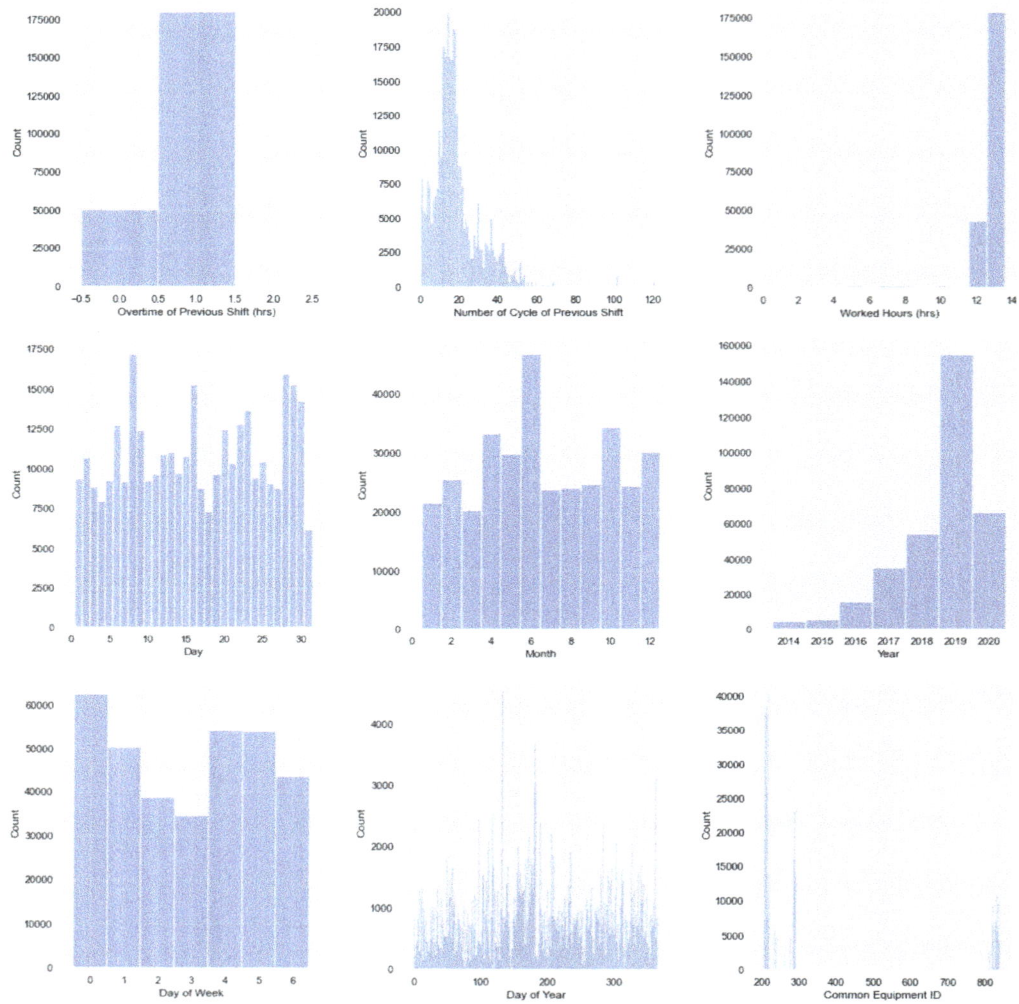

Figure 1. Distribution of the variables.

	Worked_Hour	Overtime	Overtime_Pre_Shift	Number of Cycle	Number of Cycle_Prev_Shift	Work Type_Unscheduled	Status_Ready	Status_Delay	Reason Description_PRODUCTION
Worked_Hour	1.000	0.556	0.337	0.035	0.086	-0.005	0.000	0.002	-0.001
Overtime	0.556	1.000	0.825	0.025	0.086	-0.005	0.005	-0.006	0.003
Overtime_Pre_Shift	0.337	0.825	1.000	0.037	0.063	-0.051	0.030	-0.042	0.016
Number of Cycle	0.035	0.025	0.037	1.000	0.336	-0.036	0.036	-0.032	0.036
Number of Cycle_Prev_Shift	0.086	0.086	0.063	0.336	1.000	-0.025	0.023	-0.021	0.034
Work Type_Unscheduled	-0.005	-0.005	-0.051	-0.036	-0.025	1.000	-0.688	0.714	-0.684
Status_Ready	0.000	0.005	0.030	0.036	0.023	-0.688	1.000	-0.839	0.995
Status_Delay	0.002	-0.006	-0.042	-0.032	-0.021	0.714	-0.839	1.000	-0.835
Reason Description_PRODUCTION	-0.001	0.003	0.016	0.036	0.034	-0.684	0.995	-0.835	1.000

Figure 2. Correlation of the variables.

4. Methodology

Over the past several years, the UMODEL lab at University of Utah's mining engineering department has been studying and modeling mine workforce fatigue. The approach has been examining fatigue through direct surveys of crews, developing tracking technology, and modeling fatigue using operational technology. This study investigates a model of the fatigue of individuals based on job demands and environmental factors. Therefore, it uses a machine learning (ML) algorithm applied to data from a surface mine to identify indicators of fatigue in operational datasets. The process and steps of this study are provided in Figure 3.

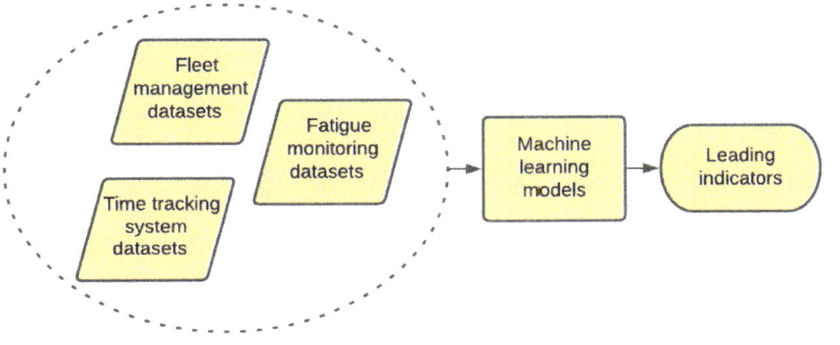

Figure 3. Process of study.

4.1. Modeling Approach

In general, there are three types of Machine Learning (ML) algorithms: supervised learning, unsupervised learning, and reinforcement learning. This study involves supervised learning, which includes a target variable (dependent variable) and a given set of predictors (independent variables or features) [25]. Dependent variables are predicted by the independent variables. An algorithm that maps inputs to desired outputs will be made with these independent variables. After the model is created, the training process continues until there is a satisfactory level of accuracy in the model [25]. Some of the examples of supervised learning consist of regression, decision tree, Random Forest (RF), K-Nearest Neighbour (KNN), logistic regression, etc. The algorithm used in this study is called eXtreme Gradient Boosting (XGBoost), which is an optimized distributed gradient boosting library [25]. It is designed to be highly efficient, flexible, and portable. It accomplishes machine learning algorithms under the Gradient Boosting framework. Gradient boosting gives a prediction model in the form of an ensemble of weak prediction models, which are typically decision trees. XGBoost algorithm provides a parallel tree boosting that can run them at the same time and solve many data science problems fast and accurately.

The fundamental idea of boosting is to integrate hundreds of simple trees with low accuracy to make a more accurate model. Every iteration will generate a new tree for the model. There are thousands of methods to generate a new tree. A common method is called the Gradient Boosting Machine [25]. It uses gradient descent to make the new tree based on previous trees. For this purpose, the objective function should be derived toward the minimum gradient direction.

XGBoost Algorithm

XGBoost model is a learning framework based on Boosting Tree models. XGBoost is based on gradient boosted decision trees designed for speed and performance. It has a strong expansion and flexibility and integrates multiple tree models to build a stronger ML model. Additionally, XGBoost uses a variety of methods to avoid overfitting [26].

In this study, the following parameters, known as hyperparameters, were adjusted to make the XGBoost model perform at its best:

1. n_estimators: is the number of iterations in training. A very small n_estimators can result in underfitting, which diminishes the learning ability of the model. However, very large n_estimators will cause overfitting, which is not good either [26].
2. min_child_weight: identifies the summation of sample weight of the smallest leaf nodes to prevent overfitting.
3. max_depth: is the maximum depth of the tree. The bigger depth of the tree makes the tree model more complex and the fitting ability stronger. However, the model is more likely to overfit.
4. subsample: is the sampling rate of all training samples.
5. colsample_bytree: is the feature sampling rate when constructing each tree. In this task, this is equivalent to the sampling rate of the landmark gene.
6. learning_rate: is a tuning parameter in an algorithm that defines the weight at each step while moving toward a minimum of a loss function. It is a very important parameter that needs to be adjusted in every algorithm. It greatly affects the model performance. To make the model more robust, we can decrease the weight of each step.

4.2. Model

For this study, different iterations of the model were conducted using available data subsets as dependent variables. The machine learning procedure diagram is displayed in Figure 4. All of the features (independent variables) were created from the available data sets for each individual. The dependent variable predicts true fatigue events from non-fatigue events. True fatigue events are classified as positive and non-fatigue events are considered negative. All independent variables in these models, also known as features, are representations of some of the mine's operation data sets. These features contain values such as the status of the haul truck in the operation cycle, over time, and the working hours of the operator (see Table 3).

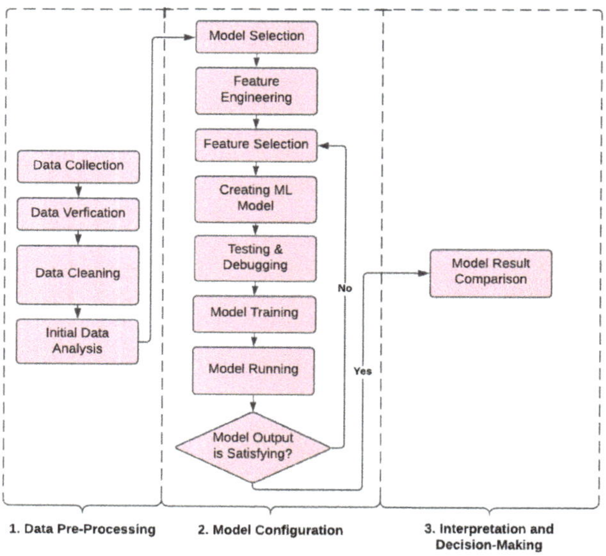

Figure 4. Machine learning model procedure diagram to predict fatigue.

Data were divided into two sets: 80% constituted the training data set, and 20% constituted the validation data set. The purpose of these models was to determine the features that can predict fatigue to maximize model scores. In these models, only data subsets with micro-sleeps were modeled. From 209,710 possible events, only 1073 contained micro-sleep with critical and low fatigue reviews in the data sets to train and validate the XGBoost algorithm. After initial data exploratory analysis, this study refined models to predict operator fatigue. Then, possible iterations to choose the best features were conducted to predict fatigue and the possibility of including all available feature sets that drive fatigue. Data for these models were constrained to the number of days contained in the fatigue data. Thus, the models were created using data from 7 November 2014 to 23 June 2020.

After engineering data, data should be numeric for applying the selected algorithm. Therefore, every feature is checked, and its format is changed to be numeric. All of the categorical features are used to make several features with 0 and 1 values as they can be appropriate for this model. The status events and production datasets create over fifty numerical variables, including status, reason description, work type, and the state of the haul truck in the time of fatigue.

In order to get the best possible results out of the selected algorithm and leverage the maximum power of the algorithm, hyperparameters should be tuned. This selected algorithm provides an extensive range of hyperparameters. XGBoost is a powerful machine learning (ML) algorithm that has shown strong performance at picking up patterns in the data by automatically tuning thousands of learnable parameters. In tree-based models, like XGBoost, the learnable parameters are the choice of decision variables at each node, creating more design decisions and, as a result, a wider range of hyperparameters. These parameters were specified by hand to the algorithm and fixed throughout a training phase. As mentioned earlier, for this model, there are hyperparameters including maximum depth of the tree, number of trees to grow, number of variables to consider when building each tree, minimum number of samples on a leaf and fraction of observations used to build a tree. These are some of the model parameters used in this study: learning rate: 0.05, maximum depth of tree: 5, number of trees: 100, minimum child weight: 500, fraction of observations used to build a tree (subsample): 1.

4.2.1. Model Iterations

For the first iteration of the modeling, all the available features were used to model fatigue. Variables for the second, third and fourth iterations of modeling are determined by results from the first iteration. Two top features from the first model iteration are the number of cycles and overtime of the employees. In order to see the effect of other features, they are removed for the second iteration of the modeling. Another top feature of the first model was employee ID. It shows that some individuals have a higher rate of fatigue compared to others. Therefore, two different models were created based on the employee ID: employees with higher rates of fatigue and employees with lower rates of fatigue. Results from both models demonstrate that different indicators affect fatigue of these two groups.

4.2.2. Model Evaluation

After data training and model debugging, each model result was interpreted, and if the result is accepted, the model is evaluated. Different approaches are available for evaluating the model. One way of the model evaluation is the model score or R^2. This score is usually from the validation data set. The higher the model score, the better the model performs. Next for each tree-based algorithm is Gini index. The Gini index calculates the degree of probability of a specific variable that is wrongly being classified when chosen randomly for each tree, which works on categorical variables. The degree of Gini index varies from 0 to 1:

- Where 0 describes that all the elements be allied to a certain class.

- The Gini index of value as 1 denotes that all the elements are randomly distributed across various classes.
- A value of 0.5 shows the elements are uniformly distributed into some classes.

The next method of model evaluation, which is used here, is the confusion matrix. Since this model used classification predictions, there are four types of outcomes that could occur, which are often plotted on a confusion matrix as an outcome of the model.

- True positives: when the model predicts a fatigue event which is an actual fatigue event in the data sets.
- True negatives: when the model predicts that an event is not fatigue and it is not an actual fatigue event in the data sets.
- False positives: when the model predicts a fatigue event that is not an actual fatigue event in the data sets.
- False negatives: when the model predicts that an event is not fatigue and it is an actual fatigue event in the data sets.

5. Results

5.1. Model Results

Different iterations of the model were created, and five of them were selected as they performed better. All of these model iterations work well as their scores are acceptable. The first model with all of the features works great. It shows that employee ID, event duration, worked hour, number of cycles from the previous shift, shift index, day, day of year, day of week, and overtime of the previous shift have the biggest effect on the fatigue of the haul truck operators. It shows that some individuals have a higher rate of fatigue compared to others, as the top feature of the model shows. Crews seem to have outliers that are the main drivers of fatigue events [6]. Another parameter from this model is overtime from the previous shift, which denotes more fatigue happened for employees who have more overtime from the previous shift. Moreover, it demonstrates that the state of the haul truck categorized to such as empty-driving and full-driving can drive fatigue of the operator. All the top features are displayed in Table 4.

For the second iteration of the model, the number of cycles and overtime of the employee from the previous shift substitutes with the number of cycles and overtime of the same shift. The same result as the previous model shows that employee ID has the highest effect on fatigue. Some of the other top features of this model are shift index, event duration, number of cycles, common equipment ID, worked hour, day of year, week, work type unscheduled, and shovel machine.

After analyzing two first model outcomes, to see the effect of the top features on the model prediction, some of them are removed from the data set to run the third step of the modeling. Hence, the third model was created with all features except the number of cycles and overtime of the previous shift and the same shift. The top feature of the third model is employee ID, which shows the same information as the two first models. Other topmost features are shift index, event duration, worked hours, and equipment ID. It shows that some specific fleets have a higher rate of fatigue compared to others. Overall, it shows other important features from the model like day, week, work type, and state of the haul truck are more effective on fatigue.

A top feature of the first three models is employee ID. Therefore, to see the effect of employee ID on the model prediction, we decided to create the fourth and fifth steps of the modeling for two groups of employees with a higher and lower rate of fatigue. The fourth model also performs well for employees with a higher rate of fatigue. It shows that top features are included in the status shift index, followed by the equipment ID, event duration, day, worked hour, day of year, shovel machine, day of week. It also demonstrates that the state of haul truck of full driving has a higher effect compared to empty driving.

Another iteration of the model was conducted for the employees with a lower rate of fatigue. Model outcome shows that shift index event duration, shovel machine, day of year, and worked hours, followed by equipment ID, day, day of week, work type unscheduled, and week have effects on the fatigue. Other top features of this model are the state of haul trucks and also shows that some fleets have higher fatigue compared to others. In Table 4, the results of the best-performed model are displayed. Later, a comparison of the model iterations is presented.

Table 4. Model performance results.

Model	Independent Variables	Score	Top Features
First model	Employee ID Common Equipment ID Year Month Week Day Day of week Day of year Shift is end of month Shift is start of month Shift is end of quarter Shift is start of quarter Shift is end of year Shift is start of year Worked hour Overtime of previous shift Status—Down Status—Standby Number of cycles of previous shift Work Type Scheduled Work type Unscheduled Fleet_HTE—KOM960 Fleet_HTE—LEB282 Fleet_HTM—CAT793 Reason Description (Included of 70 different variables) State of Truck—Empty Driving State of Truck—Full Driving State of Truck—Loading State of Truck—None State of Truck—In Queue Waiting for Loading State of Truck—Spotting for Load Status Event Duration	0.98	1. Employee ID 2. Event Duration 3. Worked Hour 4. Number of Cycle Previous Shift 5. Shift Index 6. Day 7. Day of year 8. Day of week 9. Overtime Previous Shift 10. Common Equipment ID 11. Is Quarter End 12. Shovel Machine 13. Month 14. Week 15. State Truck Full Driving 16. State Truck Empty Driving
Second model	Employee ID Common Equipment ID Year Month Week Day Day of week Day of year Shift is end of month Shift is start of month Shift is end of quarter Shift is start of quarter Shift is end of year Shift is start of year Worked hour Overtime Status—Down Status—Standby Number of cycles Work Type Scheduled Work type Unscheduled Fleet_HTE—KOM960 Fleet_HTE—LEB282 Fleet_HTM—CAT793 Reason Description (Included of 70 different variables) State of Truck—Empty Driving State of Truck—Full Driving State of Truck—Loading State of Truck—None State of Truck—In Queue Waiting for Loading State of Truck—Spotting for Load Status Event Duration	0.83	1. Employee ID 2. Shift Index 3. Event Duration 4. Number of Cycle 5. Common Equipment ID 6. Worked Hour 7. Day of Year 8. Week 9. Work Type Unscheduled 10. Shovel Machine 11. Work Type Scheduled 12. Day of Week 13. Day 14. State Truck Full Driving 15. State Truck Empty Driving 16. Fleet HTE—KOM960 17. State Truck Queue Waiting for Load 18. Fleet HTE—CAT793 19. Fleet HTE—LEB282 20. Year

Table 4. *Cont.*

Model	Independent Variables	Score	Top Features
Third model	Employee ID Common Equipment ID Year Month Week Day Day of week Day of year Shift is end of month Shift is start of month Shift is end of quarter Shift is start of quarter Shift is end of year Shift is start of year Worked hour Status—Down Status—Standby Work Type Scheduled Work type Unscheduled Fleet_HTE—KOM960 Fleet_HTE—LEB282 Fleet_HTM—CAT793 Reason Description (Included of 70 different variables) State of Truck—Empty Driving State of Truck—Full Driving State of Truck—Loading State of Truck—None State of Truck—In Queue Waiting for Loading State of Truck—Spotting for Load Status Event Duration	0.82	1. Employee ID 2. Shift Index 3. Event Duration 4. Worked Hour 5. Common Equipment ID 6. Day of Year 7. Week 8. Shovel Machine 9. Day 10. Day of Week 11. Work Type Unscheduled 12. Work Type Scheduled 13. State Truck Full Driving 14. Fleet HTE—KOM960 15. State Truck Empty Driving 16. State Truck Queue Waiting for Load 17. Year 18. Fleet HTE—LEB282 19. Fleet HTE—CAT793 20. Month
Fourth model (With employees with a higher rate of fatigue)	Common Equipment ID Year Month Week Day Day of week Day of year Shift is end of month Shift is start of month Shift is end of quarter Shift is start of quarter Shift is end of year Shift is start of year Worked hour Status—Down Status—Standby Work Type Scheduled Work type Unscheduled Fleet_HTE—KOM960 Fleet_HTE—LEB282 Fleet_HTM—CAT793 Reason Description (Included of 70 different variables) State of Truck—Empty Driving State of Truck—Full Driving State of Truck—Loading State of Truck—None State of Truck—In Queue Waiting for Loading State of Truck—Spotting for Load Status Event Duration	0.83	1. Shift Index 2. Common Equipment ID 3. Event Duration 4. Day 5. Worked Hour 6. Day of year 7. Shovel Machine 8. Day of Week 9. State Truck Full Driving 10. Week 11. State Truck Empty Driving 12. Year 13. Month 14. State Truck Queue Waiting for Load 15. Fleet HTE—KOM960 16. Work Type Unscheduled 17. Is Month End

Table 4. *Cont.*

Model	Independent Variables	Score	Top Features
Fifth model (With employees with a lower rate of fatigue)	Common Equipment ID Year Month Week Day Day of week Day of year Shift is end of month Shift is start of month Shift is end of quarter Shift is start of quarter Shift is end of year Shift is start of year Worked hour Status—Down Status—Standby Work Type Scheduled Work type Unscheduled Fleet_HTE—KOM960 Fleet_HTE—LEB282 Fleet_HTM—CAT793 Reason Description (Included of 70 different variables) State of Truck—Empty Driving State of Truck—Full Driving State of Truck—Loading State of Truck—None State of Truck—In Queue Waiting for Loading State of Truck—Spotting for Load Status Event Duration	0.82	1. Shift Index 2. Event Duration 3. Shovel Machine 4. Day of year 5. Worked Hour 6. Common Equipment ID 7. Day 8. Day of Week 9. Work Type Unscheduled 10. Week 11. Work Type Scheduled 12. State Truck Full Driving 13. State Truck Empty Driving 14. Fleet HTE—KOM960 15. Fleet HTE—LEB282 16. State Truck Loading 17. State Truck Queue Waiting for Load 18. Month 19. Fleet HTE—CAT793 20. Is Month End

All the model outputs illustrate that the state of the haul truck can increase the fatigue of the operators. Among all the different states of the haul truck, empty driving has a greater impact, which is not surprising since the truck needs to be moving for the system to work. Another observation from the models is that, after empty driving, full driving also has an effect on the fatigue of the operators. It may be because of the monotonous task while driving to the destination of dumping of loading. It also may be due to long-distance driving.

5.2. Gini Index

The Gini index or Gini coefficient computes the degree of probability of a specific variable that is wrongly being classified when chosen randomly. The Gini index estimates the amount of probability of a specific feature that is incorrectly classified when selected randomly in the decision tree. If all the elements are linked with a single class, then it can be called pure. The Gini index varies between values 0 and 1, where 0 identifies the purity of classification, and all the elements belong to a specified class or only one class exists there, and 1 expresses the random distribution of elements across different classes. Additionally, the value of 0.5 shows an equal distribution of elements over some classes. In every decision tree algorithm, Gini index can help to find the best-chosen samples for the best-performed tree. The best-chosen samples for the decision tree for each iteration of the models are shown in Table 5. Gini index is the evaluation method during the process of the model training; however, a confusion matrix is calculated after the model is made.

Table 5. Gini index of the models.

Model	Gini Index
First model	0
Second model	0.46
Third model	0.48
Fourth model	0.37
Fifth model	0.47

5.3. Confusion Matrix

Confusion matrices demonstrate counts from predicted and actual values. It shows the score of the model and how accurate the model predicted testing data. In the confusion matrix, there are four different values. The output TN means True Negative, which shows the number of negative samples classified by the model accurately. Likewise, TP stands for True Positive, indicating the number of positive samples predicted by the model accurately. FP shows False Positive values, the number of actual negative examples classified by the model as positive. FN stands for False Negatives value, which is the number of actual positive samples classified by the model as negative. Figures 5–9 represent the confusion matrices of all four model iterations. One of the commonly used metrics for classification models is accuracy. The accuracy of the model can be calculated by summation of accurate prediction over a summation of all classified samples in the confusion matrix, as Equation (1) shows. Table 6 represents the accuracy of the models.

$$\left(\frac{\mathbf{TN}+\mathbf{TP}}{\mathbf{TN}+\mathbf{TP}+\mathbf{FP}+\mathbf{FN}}\right), \quad (1)$$

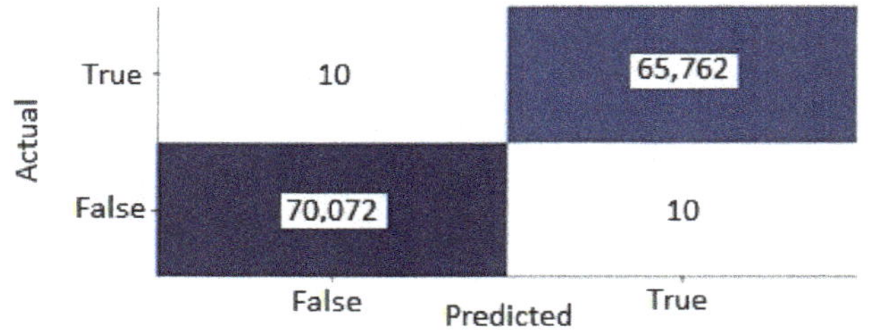

Figure 5. Confusion matrix of the first model.

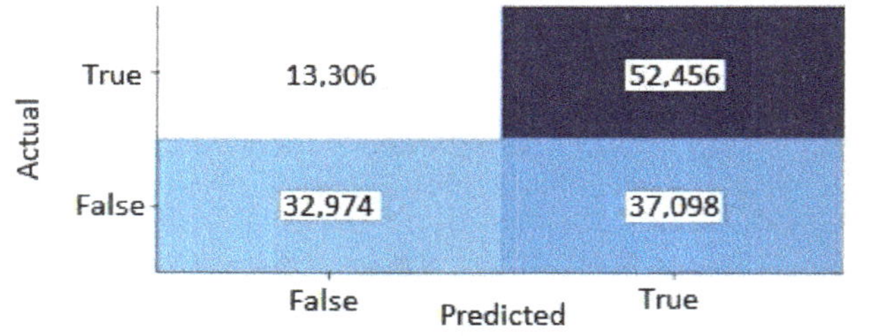

Figure 6. Confusion matrix of the second model.

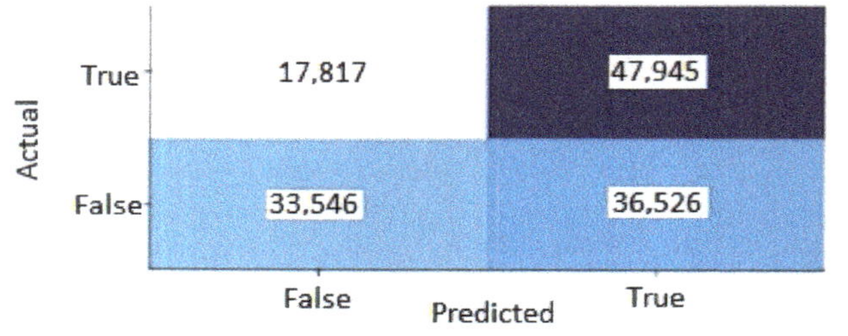

Figure 7. Confusion matrix of the third model.

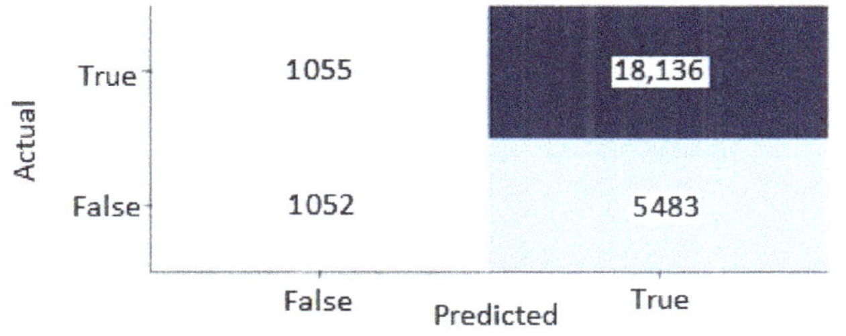

Figure 8. Confusion matrix of the fourth model.

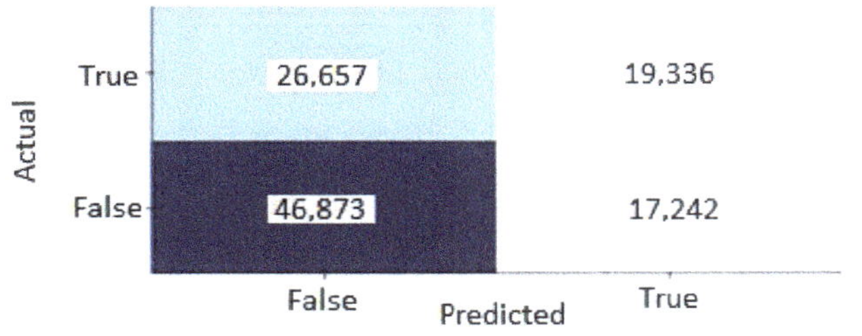

Figure 9. Confusion matrix of the fifth model.

Table 6. Model accuracy.

Model	Accuracy
First model	0.99
Second model	0.63
Third model	0.60
Fourth model	0.74
Fifth model	0.60

5.4. SHAP Values of the Models

SHAP values are based on Shapley values, a concept coming from game theory. This game theory requires a game and some players. Here, in the machine learning model, the game reproduces the outcome of the model, and the players are the features included in the model. Shapley quantifies the contribution of each player to the game, and the contribution of each feature brings to the prediction of the model. In fact, SHAP is about the local interpretability of a predictive model. Therefore, SHAP values of the five different iterations of the models are provided in Figures 10–14.

They show the feature value on the model and the SHAP value of each value of the features. Red presents the higher value of the feature, and blue presents the lower value of the feature. For instance, the employee ID with the higher value positively impacts the model output. Adversely, unscheduled work type negatively impacts the fatigue model output, which means that a higher value of the unscheduled work type increases the fatigue of the operator. Another interesting finding from the SHAP value plot is that in the case of driving with a full haul truck, a higher value has a negative effect on the fatigue model output. Therefore, these plots can be utilized to interpret the result of the model in detail and in a more nuanced way.

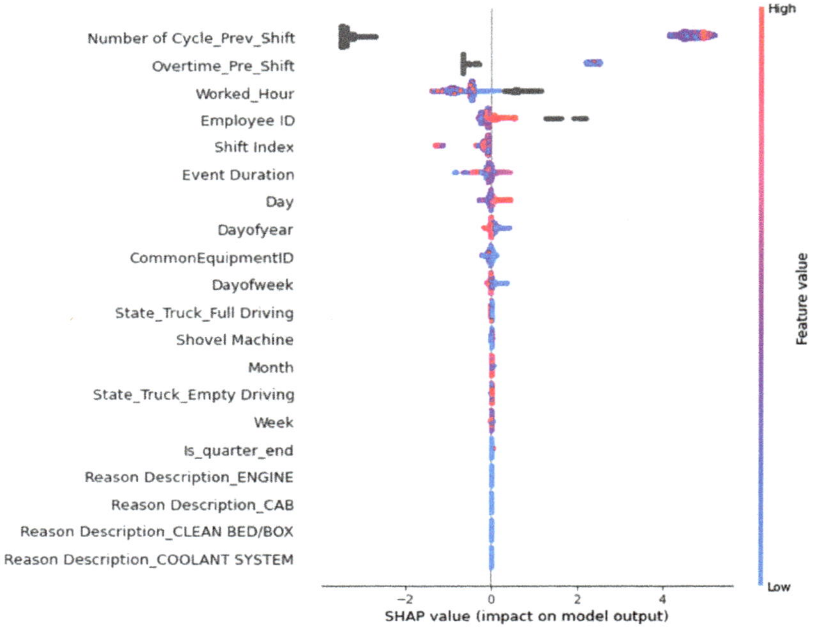

Figure 10. SHAP value of the first model.

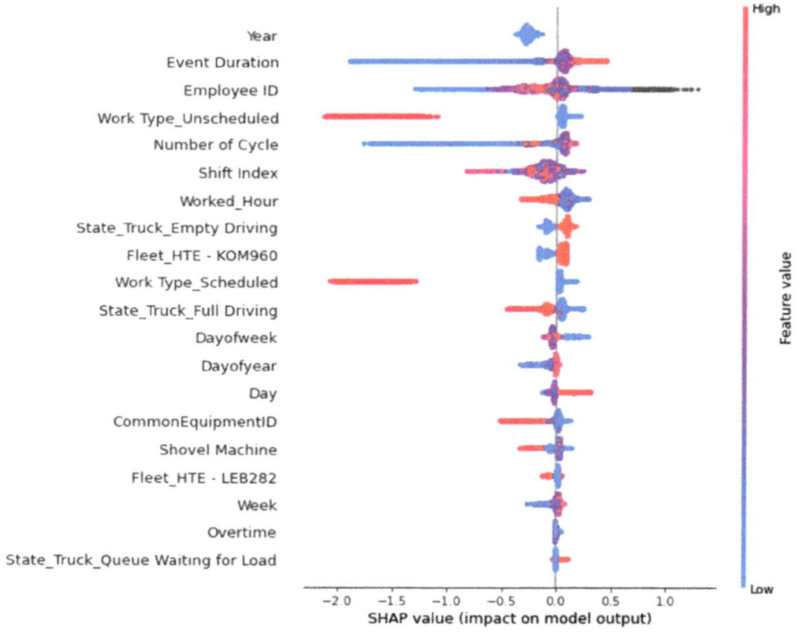

Figure 11. SHAP value of the second model.

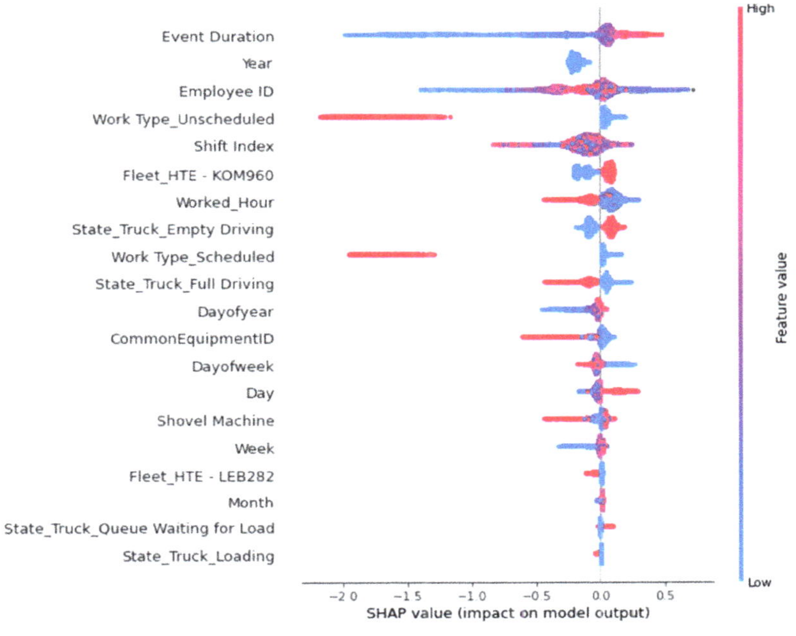

Figure 12. SHAP value of the third model.

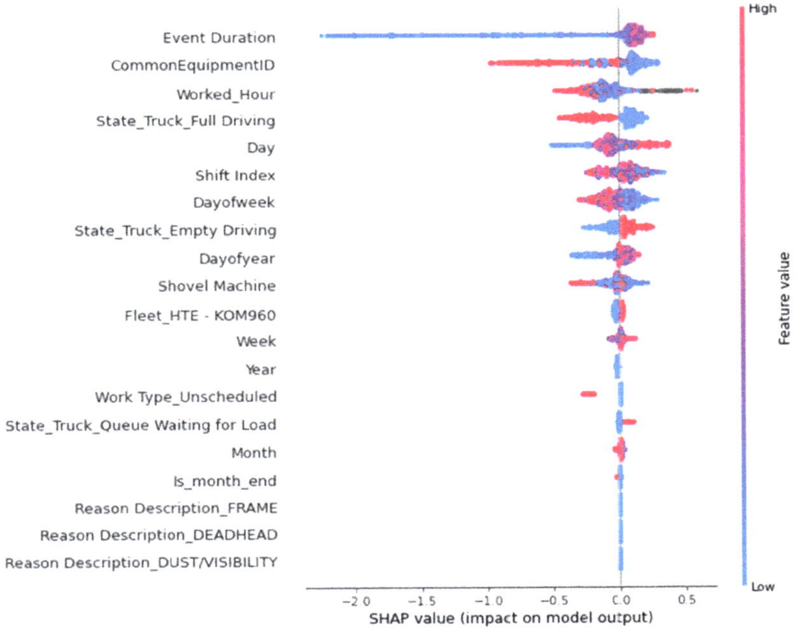

Figure 13. SHAP value of the fourth model.

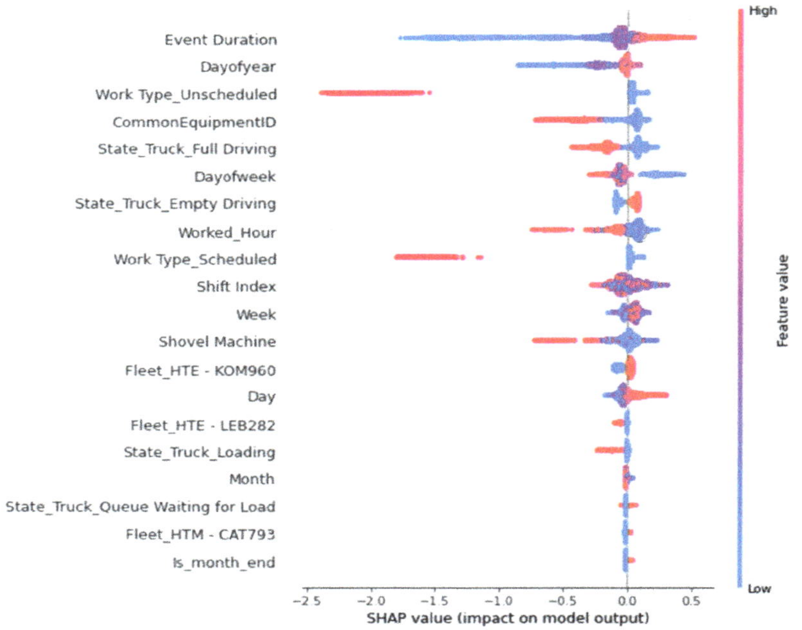

Figure 14. SHAP value of the fifth model.

5.5. Model Iterations Comparison

As discussed previously, several different model iterations were created to explore which features have the higher effects on fatigue. Five different iterations are provided for comparison. Details of them are provided in Table 7 for comparison. For the first iteration, all of the available engineered features such as the number of cycles and overtime of the previous shift are used. This model has a score of 0.98. Figure 4 displays the confusion matrix of the models, which shows that the model works decently. As it illustrates, only 20 samples are not predicted correctly. It shows that employee ID has the highest effect on fatigue. Employee ID illustrates that each individual has a different rate of fatigue events. Moreover, it shows employee ID, event duration, worked hours, and shift index has effects on the fatigue of the individuals. It also shows that the number of cycles of the previous shift has an effect on fatigue. For the second iteration, the same features are used, except the number of cycles and overtime from the previous shift are dropped from the model, and instead, those features from the same shift are used. However, the score of the model is lower, and it demonstrates the same top features as the first model, with employee ID having the highest effect on the model.

Table 7. Comparison of the models.

Model	Top Feature	Score	Percentage of Samples That Predicted Correctly
First model	Employee ID	0.99	100%
Second model	Employee ID	0.83	63%
Third model	Employee ID	0.82	60%
Fourth model	Shift Index	0.83	74%
Fifth model	Shift Index	0.82	60%

For the third model iteration, employee ID and for the fourth and fifth models, shift index is a top feature. The second, third, fourth, and fifth models are not performing well as the first model, but they demonstrate other important features of the model. For example, event duration, shovel machine, equipment ID, and day are the features that model outcome shows as top features. The confusion matrix for these models shows that the model has some errors in predicting fatigue in these models. The second model has a significant error in predicting samples, 13,306 and 37,098, respectively true and false samples are predicted wrong by the model. The third model could not predict 17,817 true samples and 36,526 false samples correctly. The fourth model also predicted 1055 and 5483 true and false samples wrong. In the last model, 26,657 and 17,242 true and false samples are predicted wrong. As the confusion matrix represents, fatigue is predicted by the second, third, and fourth models mostly when a sample is a fatigue event, which means they are positive samples. However, the fifth model predicts better when the sample is not a fatigue event, which means that samples are negative samples.

6. Discussion

The model output identifies the variables that have the greatest impact on all fatigue events. Table 8 illustrates the most important features and their data sources from the best-performed model (first model). The model results admit current understanding of fatigue, at the same time providing some interesting new insights into work and environmental factors that potentially cause fatigue for individuals. Fatigue events are clustered consistently within a group of individuals. Since the model outcome represents employee ID as one of the top factors, we can conclude that individual factors greatly affect fatigue. Based on the study by Drews F. (2020), this can be because of different factors like individual sleep efficiency, clinical conditions, life and event stressors, and personality factors [18]. From the model outcome and as it is expected, each individual has a different rate of fatigue.

Table 8. Top features by data classification.

Data Category	Feature Rank	Feature
Time and attendance	1	Employee ID
	3	Worked Hour
	5	Shift index
	6	Day
	7	Day of year
	8	Day of Week
	9	Overtime Previous Shift
	11	Is quarter end
	13	Month
	14	Week
Fleet management system (production and status)	2	Event Duration
	4	Number of Cycle Previous Shift
	10	Common Equipment ID
	12	Shovel Machine
	15	State Truck Full Driving
	16	State Truck Empty Driving

Drews F. (2020) conceptual fatigue model examines previous shift factors that have effects on the sleep history and, finally, fatigue of the individual [18]. It also represents work demand as a big factor of fatigue, which has physiological and psychological impacts on the individual. A similar result from this study outcome shows that overtime work and the number of cycles of the previous shift highly impact fatigue. The number of cycles and overtime show the burden of the work demand for the operators. Similarly, the state of the haul truck driver is another factor that drives fatigue. They demonstrated the state of the haul truck in the load-dump cycle when fatigue happened. Another finding from these models shows that the shift index is a factor that drives fatigue, which shows fatigue rate is higher in some specific shifts. Additionally, the outcome from the models shows that full-driving and empty-driving have a higher fatigue rate than other states, which full-driving affects fatigue more compared to empty-driving. It can be because of the monotonous task for a long time compared to when they dumped, loaded, or waited in a queue. Moreover, model results offer that some specific work types, like unscheduled ones, increase the rate of fatigue. It implies that any unscheduled tasks like delays in the cycles make the operator more vulnerable to fatigue due to waiting time. It also shows that cycles after that will have a higher risk of fatigue. Other variables from the model are shovel machine, equipment ID, day, week, month, and is the end of the month, which suggest a pattern in the fatigue time for individuals. As the model shows, the higher duration of the fatigue event, the more fatigue event happens for the operator, which shows a more serious issue.

Results from the model with higher rates of fatigue and lower rates of fatigue demonstrate that different indicators affect the fatigue of these two groups. In addition, they display that some fleets have a higher rate of fatigue compared to other fleets. These models also show that unscheduled work type has a higher impact on the fatigue of the employees with a higher rate of fatigue.

These outcomes can help the health and safety managers understand the magnitude of the mine site's fatigue issues. Looking at the significant effect of the individual factors on fatigue and work environment factors propose more attention to individuals by the health and safety managers. The model can be used to justify targeted fatigue training for each individual that has a higher fatigue risk to take care of their individual factors like sleep quantity and quality. Another approach would be providing insight to managers and supervisors to target more flexible interventions (shift schedule, breaks, etc.) for individuals with a higher rate of fatigue or a greater fatigue duration. Supervisors can have more targeted engagement with operators during monotonous tasks like empty haul state. The number of cycles shows if they worked the whole shift or had some equipment downtime. The high number of cycles shows high work-demand during the shift. Similarly, overtime shows the burden of the work, which even asks for work after the shift ends. From the health and safety perspective, they can manage to support and check individuals with a lower rate of break. Another issue is the delay in production, which supervisors can manage by being alert to check these operators more often.

These model outcomes proved that factors that drive fatigue for each individual are different, and the mining industry needs to have individualized flexibility of health and safety programs versus a common general program or a tool to detect fatigue. Current fatigue monitoring systems are not able to consider these individual differences in a comprehensive way. A more comprehensive fatigue monitoring and prediction program can likely prevent the consequences earlier than lagging systems. Looking at the individual's condition is very important and helpful in improving health and safety situations. Moreover, work demand is another factor that health and safety programs could look at to control fatigue. Such as having specific controls and supervision in a time of higher work demand.

7. Conclusions

Although this study tries to show the application of machine learning algorithms in health and safety management mining operations, its finding helps to understand the individual's fatigue. This finding, along with a previous study from the authors, confirms that fatigue is caused by a wide variety of individual and work environmental factors. Some of them are easy to quantify, and some are difficult. Since fatigue is the complex interaction between human behavior and the dynamic work mines environment, it is tough to make a comprehensive model that shows all of the variables driving the fatigue of each individual. Previous models examined the issues at an aggregated level using operational data sets; this study clearly shows individualized factors from the operational data sets that have effects on the individual's fatigue.

As it is mentioned before, JDR model is related to risk factors associated with job stress, such as job demands and job resources [8]. Our model uses these job demands factors that could contribute to physical or physiological stressors for the operators like the number of cycles, overtime, worked hours, and other production variables, etc. However, this model is limited by the available variables from the data sets, it would be helpful to add other job-related factors for the next study, like off days and any break time for the operators. In addition, other personal factors like sleep duration, efficiency, exercise, food and drink consumption would also aid in developing a more comprehensive understanding of fatigue risk for individuals.

All developed models have a high score greater than 0.8, but the first iteration has the highest score by far. However, this model is used for guiding other iterations of the additional models. These subsequent models did not achieve as high a score as the initial model. Findings of the first model show that fatigue is clustered around certain Individual's and factors from the previous shift are very important. The important point is how to find these factors from the available data. It is recommended for the next research to use individual factors like fitness, sleep history, commute hours, diet, and other individual factors to explore more possible indicators of fatigue. Another recommendation is to use the Neural Network model to understand the combination of the parameters that have effects on the individual's fatigue.

Author Contributions: Conceptualization, W.P.R., E.T. and F.A.D.; methodology, E.T.; software, E.T.; validation, W.P.R., E.T. and F.A.D.; formal analysis, E.T.; investigation, E.T.; resources, W.P.R.; data curation, E.T.; writing—original draft preparation, E.T.; writing—review and editing, W.P.R. and E.T.; visualization, E.T.; supervision, W.P.R.; project administration, W.P.R.; funding acquisition, W.P.R. All authors have read and agreed to the published version of the manuscript.

Funding: This research was funded by National Institute for Occupational Safety and Health (NIOSH) with grant number 75D30119C05500.

Data Availability Statement: The data presented in this study are not publicly available due to confidentiality.

Acknowledgments: We would like to express our gratitude to the National Institute for Occupational Safety and Health (NIOSH) for funding and the mining company, its management, and all of the people supporting this research.

Conflicts of Interest: On behalf of all authors, the corresponding author states that there is no conflict of interest.

References

1. Sadeghniiat-Haghighi, K.; Yazdi, Z. Fatigue management in the workplace. *Ind. Psychiatry J.* **2015**, *24*, 12. [PubMed]
2. Halvani, G.H.; Zare, M.; Mirmohammadi, S.J. The relation between shift work, sleepiness, fatigue and accidents in Iranian Industrial Mining Group workers. *Ind. Health* **2009**, *47*, 134–138. [CrossRef] [PubMed]
3. Bauerle, T.; Dugdale, Z.; Poplin, G. Mineworker fatigue: A review of what we know and future directions. *Min. Eng.* **2018**, *70*, 33. [PubMed]
4. Qing, W.; BingXi, S.; Bin, X.; Junjie, Z. A perclos-based driver fatigue recognition application for smart vehicle space. In Proceeding of the 2010 Third International Symposium on Information Processing, Qingdao, China, 15–17 October 2010.
5. Wang, L.F.; Liu, J.Q.; Yang, B.; Yang, C.S. PDMS-based low cost flexible dry electrode for long-term EEG measurement. *IEEE Sens. J.* **2012**, *12*, 2898–2904. [CrossRef]
6. Prediction and Management of Fatigue with Voice Analysis. Available online: https://wombatt.net/ (accessed on 15 July 2022).
7. Bakker, A.B.; Demerouti, E. The Job Demands-Resources model: State of the art. *J. Manag. Psychol.* **2007**, *22*, 309–328. [CrossRef]
8. Bakker, A.B.; Demerouti, E.; De Boer, E.; Schaufeli, W.B. Job demands and job resources as predictors of absence duration and frequency. *J. Vocat. Behav.* **2003**, *62*, 341–356. [CrossRef]
9. Demerouti, E.; Bakker, A.B.; Nachreiner, F.; Schaufeli, W.B. The job demands-resources model of burnout. *J. Appl. Psychol.* **2001**, *86*, 499. [CrossRef] [PubMed]

10. Briggs, C.; Nolan, J.; Heiler, K. *Fitness for Duty in the Australian Mining Industry: Emerging Legal and Industrial Issues*; Australian Centre for Industrial Relations Research and Teaching: Sydney, Australia, 2001; pp. 1–52. ISBN 186487421X.
11. Parker, T.W.; Warringham, C. Fitness for work in mining: Not a 'one size fits all' approach. In Proceedings of the Queensland Mining Industry Health & Safety Conference, Casino, Australia, 4–7 August 2004.
12. Hutchinson, B. Fatigue Management in Mining–Time to Wake up and Act. Optimize Consulting. 2014. Available online: http://www.tmsconsulting.com.au/basic-fatigue-management-in-mining/ (accessed on 20 January 2020).
13. Pelders, J.; Nelson, G. Contributors to Fatigue of Mine Workers in the South African Gold and Platinum Sector. *Saf. Health Work* **2019**, *10*, 188–195. [CrossRef] [PubMed]
14. Cavuoto, L.; Megahed, F. Understanding fatigue and the implications for worker safety. In Proceedings of the ASSE Professional Development Conference and Exposition 2016, Atlanta, GA, USA, 26–29 June 2016.
15. Talebi, E.; Roghanchi, P.; Abbasi, B. Heat management in mining industry: Personal risk factors, mitigation practices, and industry actions. In Proceedings of the 17th North American Mine Ventilation Symposium, Montreal, QU, Canada, 28 April–1 May 2019.
16. Tufano, P. Who manages risk? An empirical examination of risk management practices in the gold mining company. *J. Financ.* **1996**, *51*, 1097–1137. [CrossRef]
17. Wegerich, S.W.; Wolosewicz, A.; Xu, X.; Herzog, J.P.; Pipke, R.M. Automated Model Configuration and Deployment System for Equipment Health Monitoring. U.S. Patent 7,640,145 B2, 29 December 2009.
18. Drews, F.A.; Rogers, W.P.; Talebi, E.; Lee, S. The Experience and Management of Fatigue: A Study of Mine Haulage Operators. *Min. Metall. Explor.* **2020**, *37*, 1837–1846. [CrossRef]
19. Hallowell, M.R.; Hinze, J.W.; Baud, K.C.; Wehle, A. Proactive construction safety control: Measuring, monitoring, and responding to safety leading indicators. *J. Constr. Eng. Manag.* **2013**, *139*, 04013010. [CrossRef]
20. Tixier, A.J.P.; Hallowell, M.R.; Rajagopalan, B.; Bowman, D. Application of machine learning to construction injury prediction. *Autom. Constr.* **2016**, *69*, 102–114. [CrossRef]
21. Thibaud, M.; Chi, H.; Zhou, W.; Piramuthu, S. Internet of Things (IoT) in high-risk Environment, Health and Safety (EHS) industries: A comprehensive review. *Decis. Support Syst.* **2018**, *108*, 79–95. [CrossRef]
22. Lingard, H.; Hallowell, M.; Salas, R.; Pirzadeh, P. Leading or lagging? Temporal analysis of safety indicators on a large infrastructure construction project. *Saf. Sci.* **2017**, *91*, 206–220. [CrossRef]
23. Talebi, E.; Rogers, W.P.; Morgan, T.; Drews, F.A. Modeling mine workforce fatigue: Finding leading indicators of fatigue in operational data sets. *Minerals* **2021**, *11*, 621. [CrossRef]
24. Dong, G.; Liu, H. (Eds.) *Feature Engineering for Machine Learning and Data Analytics*; CRC Press: Boca Raton, FL, USA, 2018; pp. 2–9.
25. Zhang, L.; Zhan, C. Machine learning in rock facies classification: An application of XGBoost. In Proceedings of the International Geophysical Conference, Qingdao, China, 17–20 April 2017.
26. Li, W.; Yin, Y.; Quan, X.; Zhang, H. Gene expression value prediction based on XGBoost algorithm. *Front. Genet.* **2019**, *10*, 1077. [CrossRef] [PubMed]

mining MDPI

Review

Electrification Alternatives for Open Pit Mine Haulage

Haiming Bao *, Peter Knights, Mehmet Kizil and Micah Nehring

School of Mechanical and Mining Engineering, The University of Queensland, Brisbane, QLD 4072, Australia
* Correspondence: h.bao@uq.edu.au

Abstract: Truck-Shovel (TS) systems are the most common mining system currently used in large surface mines. They offer high productivity combined with the flexibility to be rapidly relocated and to adjust load/haul capacity and capital expenditure according to market conditions. As the world moves to decarbonise as part of the transition to net zero emission targets, it is relevant to examine options for decarbonising the haulage systems in large surface mines. In-Pit Crushing and Conveying (IPCC) systems offer a smaller environmental footprint regarding emissions, but they are associated with a number of limitations related to high initial capital expenditure, capacity limits, mine planning and inflexibility during mine operation. Among the emerging technological options, innovative Trolley Assist (TA) technology promises to reduce energy consumption for lower carbon footprint mining systems. TA systems have demonstrated outstanding potential for emission reduction from their application cases. Battery and energy recovery technology advancements are shaping the evolution of TAs from diesel-electric truck-based patterns toward purely electrified BT ones. Battery Trolley (BT) systems combined with autonomous battery-electric trucks and Energy Recovery Systems (ERSs) are novel and capable of achieving further significant emission cuts for surface mining operations associated with safety, energy saving and operational improvements. This article reviews and compares electrification alternatives for large surface mines, including IPCC, TA and BT systems. These emerging technologies provide opportunities for mining companies and associated industries to adopt zero-emission solutions and help transition to an intelligent electric mining future.

Keywords: decarbonization; IPCC; trolley assist; battery trolley; battery-electric trucks; electrification alternatives

Citation: Bao, H.; Knights, P.; Kizil, M.; Nehring, M. Electrification Alternatives for Open Pit Mine Haulage. *Mining* **2023**, 3, 1–25. https://doi.org/10.3390/mining3010001

Academic Editors: Juan M Menéndez-Aguado, Oscar Jaime Restrepo Baena and Jessica M. Smith

Received: 27 November 2022
Revised: 23 December 2022
Accepted: 30 December 2022
Published: 1 January 2023

1. Introduction

The current effects of climate change have created a worldwide consensus on the need for decarbonization [1]. In order to achieve the 2030 emission reduction task announced by governments, more specific technology measures will need to be applied to achieve measurable decarbonization. As an energy-intense sector, the mining industry is more dependent on fossil fuel energy than others. To date, Truck-Shovel (TS) systems are still the dominant open-pit mining haulage system, while In-Pit Crushing and Conveying (IPCC) has become an option to overcome long-distance transport in deep open pit mines. As a proven technology, Trolley Assist (TA) has shown excellent performance in saving diesel fuel and reducing emissions. With significant volatility in fuel markets, stricter environmental and social requirements, and the further advance of technologies, Battery Trolley (BT) systems are likely to guide an electrification revolution to create the first zero-emission truck fleet, which is a transition from the current diesel-electric trolley operation to battery-electric trolley haulage.

This paper investigates the current world energy outlook in carbon emissions, Australian mining sector emissions projection, renewable energy development, decarbonization technology trends and mining challenges to conclude that equipment electrification is a potential zero-emission direction. Except for conventional TS systems, there are several electrification alternatives for open pit haulage: IPCC systems, TA systems and future conceptual BT systems. At the time of this writing, due to hydrogen storage, infrastructure and logistic challenges, there is no hydrogen-power alternatives discussion in this paper. In these systems, conventional diesel-powered TS systems belong to a high-emission haulage alternative, while Fixed IPCC, Semi-Fixed IPCC, Semi-Mobile IPCC and TA systems belong to low-emission haulage alternatives. Furthermore, Fully mobile IPCC and BT systems belong to zero-emission haulage alternatives. This paper introduces all these haulage alternatives' configurations, operations, characters, and pros and cons. As an electrification revolution to create the first zero-emission truck fleet solution, BT systems combinate several state-of-the-art technologies, including autonomous trucks, battery-electric power drivetrains, TA and energy recovery technologies. Like IPCC systems, BT systems have various configurations according to charging methods, whether to build a battery station and energy recovery approaches on the downhill ramp, which are:

1. Dynamic charging BT systems;
2. Stationary charging BT systems;
3. Dual trolley BT systems.

According to mining haulage systems' requirements, there is a comparison between diesel TS, IPCC, TA and BT systems from perspectives such as flexibility, energy efficiency, CAPEX, OPEX, and others.

This paper introduces the background of worldwide decarbonization targets and mining challenges, and presents evaluation parameters to compare all mining haulage systems' pros and cons. It reviews conventional TS systems' operating processes and characteristics, and introduces current electrification alternatives for open pit mine haulage. This paper reviews IPCC and TA systems' advantages and disadvantages compared with the TS system and presents the conceptual BT systems' theory, operating process and configurations. Finally, it compares the parameters between diesel TS, Semi-Fixed/Mobile IPCC, Full-Mobile IPCC, TA, Dynamic Charging BT, Stationary Charging BT and Dual Trolley BT.

2. Decarbonization and Mining Challenges

2.1. World Energy Outlook in Carbon Emissions

Figure 1 shows CO_2 emissions in the World Energy Outlook. The Stated Policies Scenario (STEPS) takes account only of specific policies that are in place or have been announced by governments [2]. The Announced Pledges Case (APC) assumes that all announced national net zero pledges are achieved in full and on time, whether or not they are currently underpinned by specific policies [2]. The "implementation gap" between reported lowered emissions commitments and the regulatory frameworks and particular actions they need is highlighted by the 2.6 Gt difference in emissions between the STEPS and the APS in 2030. Pledges must be supported by robust, reliable policies and long-term strategies to become a reality [3]. In addition to underlining the need for specific policies and immediate measures necessary for long-term net-zero commitments, the divergence in trends between the APC and the STEPS demonstrates the potential effect of existing net-zero pledges. The APC shows, however, that existing net-zero pledges, even if fully achieved, fall well short of what is required to achieve net-zero global emissions by 2050 [2]. It clarifies what further steps must be taken to move beyond these proclaimed commitments and onto a path with a high probability of avoiding the worst impacts of climate change [3].

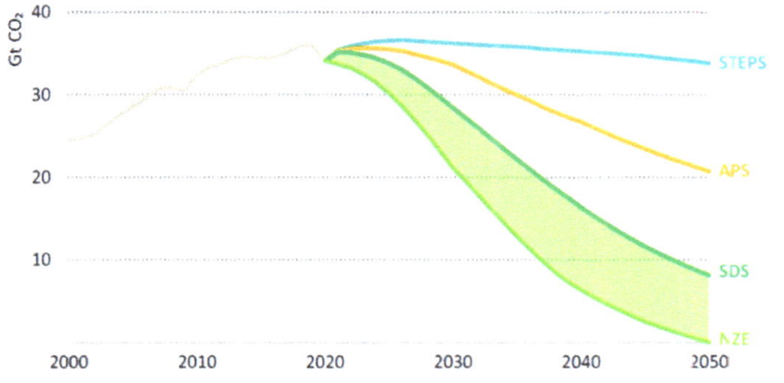

Figure 1. CO_2 emissions in the World Energy Outlook—2021 scenario over time (Source: International Energy Agency (IEA). World Energy Outlook 2021) Note: STEPS = Stated Policies Scenario; APS = Announced Pledges Scenario; SDS = Sustainable Development Scenario; NZE = Net Zero Emissions by 2050 Scenario.

2.2. *Australian Mining Sector Emissions*

Figure 2 from the Australian Department of Industry, Science, Energy and Resources indicates that from 1990 to 2020, Australian stationary energy emissions increased at an average annual rate of 1.5%. As more decarbonization measures were implemented in 2020, emissions are expected to increase more slowly, at an average rate of less than 0.1% annually. The leading causes of growing global GHG levels are emissions from transportation, electricity generation, and industrial expansion, which have pressured many industry sectors to come up with strategies to cut emissions drastically in the future. Energy efficiency, electrification equipment, and replacing fossil fuels with low-emission alternatives in the electricity generation process are essential to achieving the APC pledges, particularly over the period to 2030 [2].

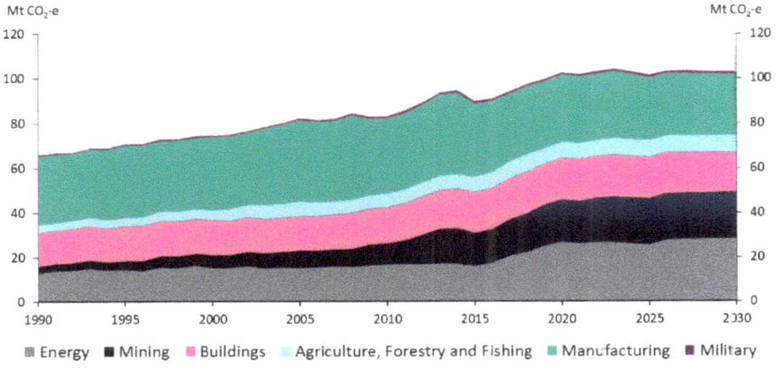

Figure 2. Australian stationary energy emissions, 1990 to 2030, Mt CO_2-e (Source: Department of Industry, Science, Energy and Resources).

Corresponding to the mining industry, a large mining base, fossil-fuel reliance, and increasing truck fleet size are the key contributors to rising CO_2 emissions [4]. The emissions from the mining subsector as whole are projected to increase from 19 Mt CO_2-e in 2020 to 21 Mt CO_2-e in 2030 because of mining needs. This increase is slowed due to technological advancements, including superior engine technology, increasing automation, and the electrification of mining equipment. Along with emissions reductions, these technological advancements also offer operating benefits such as fuel savings and productivity improvements [5].

2.3. *Renewable Energy Development*

The energy transition from fossil fuels to renewable energy resources is now one of the main challenges in achieving sustainable development goals [6]. Due to more significant widespread deployment, there have been considerable technological breakthroughs and cost reductions in wind and solar PV production during the last ten years [4]. From the most current developments in energy storage and renewable generation are reviewed, due to the growing price of fossil fuels, the adoption of carbon tax policies in some areas, and the falling capital costs of renewable generation and energy storage technology, renewables are becoming more cost-competitive [4].

A vast landmass and abundant sunshine and wind make Australia one of the world's most renewable energy-rich nations [7]. In Australia, solar and wind resources are plentiful enough to supply renewable energy needs. Figures 3 and 4 show the Australian solar and wind source maps [8]. In these figures: solar availability (good 1200–1600 MWh/m^2, very good 1600–2000 MWh/m^2, and excellent 2000–2400 MWh/m^2); wind availability (good 6–7 m/s, very good 7–8 m/s, and excellent 8–9 m/s, wind speed measured at 100 m). Solar power was the primary thermal-displacing electricity source in Queensland and South Australia. In Tasmania, a mix of wind power and hydroelectricity, and in Western Australia, New South Wales and Victoria, a mixture of solar and wind power [9].

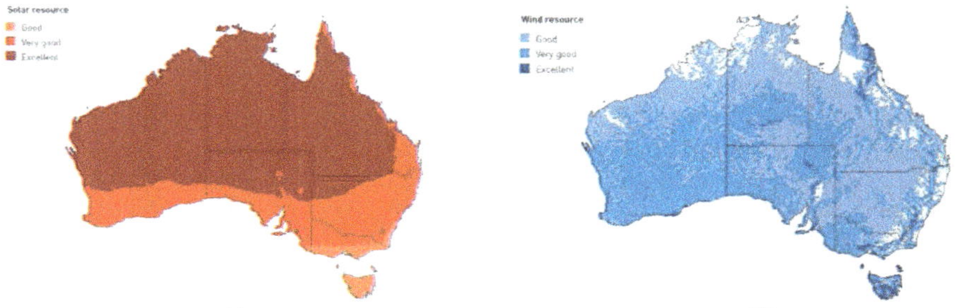

(a) **(b)**

Figure 3. (**a**) Australian solar source map, (**b**) Australian wind source map. (Source: Australian Renewable Energy Agency; The Australian Government Bureau of Meteorology Average daily solar exposure dataset; the CSIRO DATA61 Mesoscale Wind Atlas Data dataset).

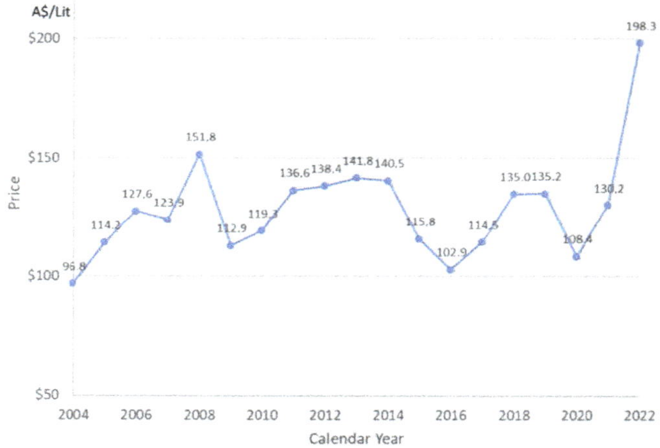

Figure 4. Historical Australia Diesel TGP Data (Source: AIP-Australian Institute of Petroleum).

Australian National Electricity Market (NEM) renewables are projected to supply over 30 percent of electricity in 2021 and 55 percent in 2030. Emissions in the NEM will decline by more than 26 percent below 2005 levels by 2022 [5].

2.4. Technology Trends

All sectors of society must support decarbonization. The mining sector must encourage the use of renewable energy technology and other cost-effective low-emission technologies as a means of combating climate change [8]. It is possible to become competitive with and replace high-emission incumbents by making the electrification of mining equipment a priority technology [5]. Through the development of electrification and battery technologies, diesel-powered equipment at mining sites and transportation may be gradually replaced with a combination of electricity-power and energy storage technology. The mining sector will likely place more emphasis on electricity generation and battery storage as a result of the switch to an "all-electric" mine [8].

Transportation is crucial in reducing emissions associated with mining operations, particularly the mining truck fleet [4]. The mining sector is replacing fossil fuels by using renewable electricity to reduce the influence of fuel price volatility and decarbonization. Especially for remote mines that rely heavily on diesel generation on-site, renewable electricity generation and zero-emission truck fleets are critical to achieving a considerable emission reduction in the total mining facility emissions [4].

2.5. Mining Challenges

With the strong demand for minerals and the depletion of high-quality resources, there are many challenges facing the mining sector. These include:

1. **Greater depths and lower grades:** Open pit mining depths have significantly expanded over the last two decades. Some open pit mines go down more than 1000 m in depth [10]. It is worth noting that future deposit extraction will inevitably be conducted at greater depths and lower grades compared to current practices, and this tendency is anticipated to continue [11,12].
2. **High operating cost:** As mines become deeper and stripping ratios increase with a lower grade, more waste material needs to be extracted. The haulage truck fleet grows correspondingly, requiring more operators and maintenance staff and a subsequent increase in diesel consumption [12–14]. In addition, as copper ore grades decline, more ore needs to be processed to attain similar metal production. A decrease in copper ore grade between 0.2% to 0.4% requires seven times more energy than present-day operations [15,16]. Reducing the cost of truck haulage, which makes up about half of the operating expenses of a mining operation, is now more essential than ever [17].
3. **Fuel price volatility:** Fossil fuel price volatility significantly impacts mining viability but is outside the control of most miners [9]. Figure 4 shows historical Australia diesel Terminal Gate Price (TGP) data. In the short term, the price of fossil fuels shows a propensity towards volatility, while it shows a significant rise from the long-term perspective.

3. Methodology

TS and IPCC systems have been widely deployed in existing open pit mines. TA systems have been proven in several mine sites and will spread to more current operating mines, while BT is still largely a conceptual decarbonization mining system, which will be put to the test in pilot mine sites. For the purpose of evaluating the selection of a mining haulage system, it is necessary to compare all these mining systems from many mining metric points, which are beneficial for mining decision-makers to select an optimum mining operating system for their mine sites. This paper adopts a mining system evaluation approach by analysing systems' operations, configurations, and characteristics to measure their reasonable implementation scopes, pros and cons from mining haulage requirements perspectives. The following is the mining haulage system evaluation important parameters:

1. Safety and productivity are indicators to measure system implementation scenarios.
2. Energy efficiency, CAPEX, OPEX, maintenance requirements, service life, additional infrastructure requirements and heat generation are system financial metrics.
3. Emissions and environmental footprint (noise/dust/DPM/vibration) are system environmental parameters.
4. Flexibility, Capacity, Scalability, Refuelling/Recharging/Swapping methods are system productivity parameters.

4. Conventional Truck-Shovel Systems

4.1. Conventional Truck-Shovel System Operating Process

Conventional TS systems continue to dominate open pit mines because diesel-powered trucks are extremely flexible in handling various materials with good grade capabilities and easy manoeuvrability [18]. The classic TS system consists of various operating processes, including manoeuvres and queues to the load point, spotting, loading, hauling the material, manoeuvres and queues to the offload point, spotting, dumping, exit tipping point, returning [19–21], which has been shown in Figure 5. To date, TS systems are the most viable, flexible and widely used mining system, and autonomous trucks have further enhanced their safety and effectiveness [19,22].

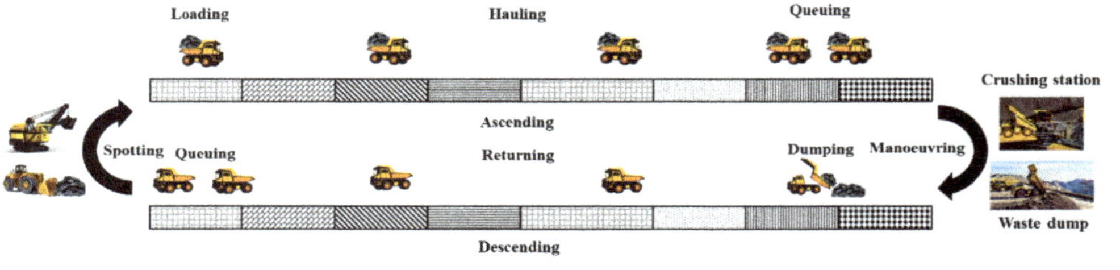

Figure 5. Conventional TS systems operating process.

4.2. Truck-Shovel Systems' Characteristics

Truck-Shovel system continues to be the predominant mining hauling system of choice for surface mines because of its ease of implementation, high flexibility, and high scalability.

1. Ease of implementation

The majority upfront expenditures of TS systems are trucks and loaders (e.g., excavators and electric shovels). A greenfield project mine may begin operations with a relatively small truck-shovel fleet and expand production capacity by purchasing larger units as the mine matures [18]. In terms of mining design and layout, several hauling segments with suitable road design, such as grade and road width, are required to complete the transport cycle. From the commencement of mines, trucks and loaders are a very predictable and controllable means of haulage from economic and operational points of view [17].

2. High flexibility

TS systems' high flexibility that they have become the dominant system in surface mining [23]. As TS systems are comprised of discrete elements, mine operators can dispatch each unit to a tailored working face to fulfil mine production requirements. Due to fewer mine design limitations and no infrastructure relocation, it's easy to change the truck fleet deployment schedule and mining tasks. Most importantly, trucks are capable of adapting to different orebody shapes and geology conditions to adapt to unforeseen changes in an open pit mine.

3. High scalability

As a mine matures, additional trucks will be added to accommodate longer haul cycles, and larger trucks and shovels will be used to take advantage of economies of scale [24].

Additionally, the number and type of trucks and loaders used can be readily adjusted to change the production rate [17].

However, there are plenty of challenges with conventional TS system:

1. High operating costs

Approximately 40–60% (depending on the combination of hauling equipment) of operating costs are attributable to haulage and material handling, notably TS operations [20,25,26]. The transportation distance starts to grow as mines become deeper, which dramatically reduces truck productivity. In order to reach the nominal mine production rate, the number of trucks in deployment needs to be increased. The demand for operating costs, including capital expenditures, fuel consumption, workforce, haul road size, maintenance, and repair service centres, all grow as the number of trucks increases [17].

2. CO_2 and diesel particular emissions

Diesel-powered trucks emit a considerable amount of gas emissions and unique toxic materials caused by diesel engines [27]. According to ABB statistics, a single mining truck emits up to 1000 tonnes of CO_2 annually. It would take 46,000 trees to absorb [28].

3. Labour force shortage

It is estimated that a conventional truck requires approximately seven people to operate (including operating and maintenance) [24]. A two 12-h-shift-day roster consists of 4.4 operators (0.4 to account for covering vacations and absences) and 2.7 maintenance workers [11], which makes it difficult for a remote mine to recruit enough people to meet its more significant truck fleet needs.

4. Fuel price volatility

Fuel energy cost is one of the most significant expenditures in the mining sector, accounting for 15 to 40% (depending on mining system components) of total mine operating costs on average [15]. The classic TS system is susceptible to the volatility of the fossil fuel market since a significant portion of its energy derives from diesel fuel. Energy consumption, chiefly by diesel-powered truck operations, is anticipated to increase further as mining activity expands and demand for clean energy transformation metals rises [15].

5. Safety risks

As the most prevalent kind of transportation equipment, haul trucks are involved in many accidents at operating mine sites, inspiring research interest in high potential incidents and serious accidents [29]. According to a report, approximately one-third of the deaths in Australian open pit mines are attributed to vehicle collisions [18].

6. Maintenance

Internal combustion trucks are complex, requiring highly skilled mechanics and high maintenance costs for diesel-based engines. Another high cost is off-highway tyres because tyre wear will be severe as increasing truck units. In the meantime, ancillary equipment (e.g., grader, water truck and dozer) are applied to keep haul road and reduce environmental footprint with a good maintenance condition to support TS system performance. The maintenance cost of a conventional TS system is a significant portion of the hauling operating expenditure.

Leveraging the economies of scale over the past two decades, mines prefer to use larger trucks to increase productivity and reduce operating costs. However, larger trucks cannot eliminate TS disadvantages, and they have other negative impacts on downstream processes (crushing and milling). High benches and larger blasting patterns make it more difficult to separate ore from waste and cause uncontrollable dilution. Consequently, a substantial portion of ore material that meets the cut-off grade requirement becomes waste or marginal due to dilution. Again, a larger proportion of oversize material would make the comminution stage of the crushing and milling process more costly. From the whole mine-to-mill perspective, as feed grade is decreased, the processing recovery will result in a greater percentage of valuable input materials being transferred into waste [17,30].

Large open pit mines have increasingly invested in Autonomous Haulage Trucks (AHTs). A significant reduction in collision risks has been achieved with AHTs, along with high levels of productivity and tire performance [31]. While autonomous technologies can mitigate operator costs and improve energy efficiency, which is a significant portion of the haulage cost, they require a higher investment compared to conventional trucks with the same capacity [32]. Even with this sophistication, including the necessary hardware and software, these AHTs cannot overcome the many problems with increased travel distance [17]. More importantly, although AHTs are more fuel efficient, they cannot achieve the actual decarbonization of the mining industry.

4.3. Truck-Shovel Systems' Energy Consumption

The science of measuring the performance (productivity and energy consumption) of mining equipment has evolved and reached maturity in terms of the truck haul cycle. During ramp climbs, Siemens (2009) [33] estimates that 70–80% percent of diesel fuel is consumed during haul operations. For an ultra-class dump truck, more than 40% of total energy is consumed to return the vehicle's mass to the ramp's top [33].

5. Electrification Alternatives for Open Pit Mine Haulage

5.1. In-Pit Crushing and Conveying Systems

5.1.1. IPCC Systems' Configurations

IPCC systems were first used in open pit mining operations in 1956 as an alternative to the classic TS haulage technique [34]. Using a continuous mining operation method often overcomes many of the drawbacks of the TS system. More specifically, compared to the conventional transportation system, it is possible to reduce the labour force, fuel consumption, and material size [14,28,35,36]. While most IPCC systems were used for coal and ore materials in the past, which is beneficial for downstream processes, it is seen as an unnecessary operating cost for overburden waste materials [18]. To date, however, IPCC systems have been increasingly introduced for stripping waste operations in response to the increasing hauling distance and stripping workload.

IPCC systems consist of crushers, in-pit conveyors (fully mobile), stationary conveyors, conveyor crossings, tripper car spreaders (waste), slewing spreaders (waste) and radial stackers (mineralized material). There are a variety of IPCC system alternatives available. In general, there are four distinct sorts of IPCC systems, each with unique characteristics. The four broad categories are: Fixed, Semi-Fixed, Semi-Mobile, and Fully Mobile systems [11,13,24,37] and each characteristic shows in Table 1.

Table 1. IPCC systems' characteristics.

IPCC Systems Type	Fixed IPCC	Semi-Fixed IPCC	Semi-Mobile IPCC	Fully Mobile IPCC
Crusher Type	Gyratory or jaw	Gyratory or jaw	Twin roll or sizer	Twin roll or sizer
Locations	Near the pit rim and crest	A strategic junction point in the pit	Near the operational level	Bench level in production
Relocations Time	Rarely or never relocated	Relocations every 3 to 5 years	Relocations every 6 to 18 months	Relocations as required to follow the shovel
Feed Systems	Shovel-Trucks	Shovel-Trucks	Shovel-Trucks and/or dozers	Shovels
Use	Deep hard rock mines—ore	Deep hard rock mines -waste or ore	Not common in deep hard rock mines -waste or ore	Not common in deep hard rock mines -waste or ore

1. Fixed In-pit Crushing and Conveying systems (F-IPCC)

In F-IPCC systems, the crusher is installed at a fixed location during the lifetime of the mine with rarely relocated, usually near the pit rim and the crest of the pit. Within the pit, the material is transported from the working face to the crusher unit using conventional truck haulage. After being crushed, the material is fed into a conveying system that moves it to either a spreader (waste material) or a stacker (mineralized material). F-IPCC system has its best application in deep, pre-existing pits, with low vertical advance rates, where a single crusher location can service the operation for an extended period.

2. Semi-Fixed In-pit Crushing and Conveying systems (SF-IPCC)

In SF-IPCC systems, the crusher is fixed at a strategic junction point in the pit stage for a certain period (usually 3 to 5 years). Truck haulage is also used within the pit to move material between the working face and the crushing unit, just like with F-IPCC. The differences are: SF-IPCC is designed to decrease the haulage distance to the crusher much more than F-IPCC, and in order to relocate SF-IPCC, the entire crusher station must be disassembled into multiple parts or modules.

3. Semi-Mobile In-pit Crushing and Conveying systems (SM-IPCC)

SM-IPCC is designed with a modular architecture to allow for the periodic movement of the crusher every 6 to 18 months as the working face deepens, where the crusher is operating near the mine working face (Figure 6). As the mine matures and increases in depth, the crusher is relocated deeper into the pit approximately every two to five benches (depending on the vertical advance rate) to maintain a short transport distance for the truck portion. Trucks feed crushing units, and dozers can directly push materials to feed crushers with a considerable cost reduction. Due to the continuous usage of trucks and the possibility of deploying the crusher at appropriate locations, SM-IPCC systems are the most easily accessible for current conventional hauling operations, which is also why SM-IPCC is the most flexible hauling system of all types of IPCC. Most importantly, by leveraging dozers or transport crawlers, the crusher can be relocated in hours without disassembling it, significantly reducing unproductive downtime.

Figure 6. Semi-Mobile In-pit Crushing and Conveying systems (Source: Sandvik Mining and Construction).

4. Fully Mobile In-pit Crushing and Conveying systems (FM-IPCC)

This system is distinguished by the loading unit dumping straight into the hopper of a fully mobile crusher that follows it (Figure 7). Once crushed, the material is transported straight from the working face to its destination through a network of conveyors. Utilizing a comprehensive continuous mining system and eliminating the requirement for truck haulage during steady state operation can dramatically save operational expenditure. However, as FMIPCC's flexibility is drastically constrained, the mine design must suit the system's requirements. In the meantime, truck haulage may still be required during each sinking phase of a mine because FMIPCC needs to be capable of completing mining tasks in complex geological conditions.

Figure 7. Full-Mobile In-pit Crushing and Conveying systems [38] (Source: McCarthy, 2013).

5.1.2. IPCC Systems' Characteristics

According to the literature review and mine site production experience, IPCC systems offer the following benefits compared to TS alternatives. The advantages of the IPCC are:

1. Operational expenditure

As a mine's activity grows, the pit deepens and the size of the waste dumps increases, leading to a longer truck haul cycle and the need for more trucks to meet production requirements. Compared to IPCC methods, truck haulage is often thought to be more costly as distance and elevation increase [33]. With savings opportunities arising from energy saving, workforce reduction, weight efficiency and maintenance, it is possible to significantly reduce material transport operating expenses (OPEX) by using an IPCC system compared to a truck haulage system. When other unit operations are considered, such as drilling, blasting, loading and ancillary services, estimates prepared at the University of Queensland put total mining costs at around 24% less in comparison to equivalent TS operations [14].

2. CO_2 emissions

IPCC systems are capable of a substantial reduction in CO_2 emissions because of fuel switching. An iron ore mine in Brazil with two installed FM-IPCC systems with a combined capacity of 7800 t/h, resulting in an expected decrease in diesel use of 60 million litres per year (ML/a), is an example of IPCC practice [11]. Reduced diesel consumption directly translates into reduced CO_2 emissions on site. In the same Brazilian mine, diesel savings of 60 ML/a equate to an approximate reduction of 130,000 t/a of CO_2. Considering that the average passenger vehicle emits approximately 3.552 t of CO_2 equivalent per year (Commonwealth of Australia, 2012), this is equivalent to taking more than 36,000 cars off the road per year [14]. The IPCC study [16] shows that when only fossil fuel-based energy was used, the CO_2 emissions per tonne of ore for the IPCC system were 67 kg CO_2 e/t as opposed to the TS system's 70 kg CO_2 e/t, a 3 kg decrease. It is possible to reduce greenhouse gas emissions by 14 kg CO_2 e/t ore using power generated from natural gas [17]. With renewable energy, e.g., solar-based and wind-based electricity, IPCC can be regarded as a decarbonization transport mining system.

3. Energy saving

Conveyor haulage naturally uses less energy per unit weight of material transported than truck haulage. Another important aspect is that conveyors use more (81%) of the consumed energy for the transportation of the payload in comparison to trucks (39%) [11]. More precisely, during a truck cycle, only 39% of the energy is used to move the payload; the other 61% is needed to move the truck's weight. Because the conveyor's upper and lower portions weigh much less than the overall amount of material for each metre of its length, just 19% of the energy used to move the material is wasted [17,39]. On the other hand, IPCC can reduce a mine's dependency on diesel fuel due to electricity-based [40].

4. Production efficiency

For the purpose of moving ore or waste to the appropriate areas, IPCC offers a continuous transportation system method, which typically improves production rates [36]. Conveyor haulage provides superior production efficiency on comprehensive metrics of assessing equipment performance, according to a comparison of the two systems (TS/IPCC) based on utilized time, operating time, and valuable operating time metrics [33]. While the truck fleet enables much higher available time and utilization time, IPCC achieves higher operating time and valuable operating time, which means higher production efficiency compared to the truck fleet [33].

5. Environmental footprint (noise and dust)

Since conveyors operate at a lower decibel level than conventional diesel-powered trucks, IPCC systems may help minimize noise pollution. Reducing the number of trucks on the road may significantly reduce the source of dust emissions, while some water will still be needed in conveyor systems to suppress dust at transfer points [11]. In other words, IPCC creates a better mining environment for the workforce from both noise and dust perspectives.

6. Maintenance

IPCC usage decreases reliance on large off-highway tyres, which account for a large portion of the truck fleet cost. Large off-highway tyre shortages will significantly impact truck fleet availability and mining production rate. In addition to reducing haul truck numbers, conveyors have been reported to reduce the need for ancillary equipment (graders, dozers and water trucks) by 25–30% [11].

7. Workforce reduction

IPCC offers more opportunities to remote open pit mines with limited labour availability and high workforce cost, in some cases, as low as one operator for each major component (crusher, conveyor, spreader/stacker), with minimal maintenance staff [24]. For example, a FM-IPCC system has a total workforce requirement of around 80 people [37], including operators and maintenance personnel. The exact staffing numbers will depend on the number and installed length of conveyors. In comparison, it is estimated that a large ultra-class mining truck requires staffing of 7 people per year. Thus, from a workforce point of view, an IPCC system becomes an attractive alternative if it can replace approximately 12 trucks [14].

8. Safety

Abbaspour [29] demonstrated that these transportation systems behave differently in terms of safety and social metrics by creating a simulation model over the whole mine life, including TS and four types of IPCC systems. In conclusion, FM-IPCC stood first in terms of safety, while the TS system was ranked fifth. A reduced truck fleet size will reduce the possibility of vehicle collisions, which is a leading cause of safety incidents in surface mining operations [14]. Because TS systems work in collaboration to feed fixed, semi-fixed and semi-mobile IPCC systems [18], FM-IPCC is able to arrive at the lowest safety risk indicator by eliminating the truck fleet.

9. Total cost operation over the mine life

In 2012, the typical capital expenditure for an IPCC system was between USD180 million and USD250 million (depending on the types of IPCC, the number was estimated to be and installed length of conveyors) [41]. For a 360-ton haul truck, the cost in 2009 was around USD5 million [42]. It can be seen from Figure 9 that IPCC requires more significant capital investment. The considerable gap causes plenty of greenfield project mines to embrace the TS system when they hope to recover capital as quickly as possible. However, the IPCC's accessories, such as the crusher and conveyor, are generally replaced every 20 to 25 years (about 150,000 h), whereas the economic replacement age for trucks is around seven to ten years of operation (between 45,000 and 60,000 h). This indicates that two truck fleet replacements will be required for a mining project that is 25 years old (i.e., where the red line drops in Figure 8). The conveyor system will require a lower total cost operation over the mine life [17]. It is estimated that the operating expenses (OPEX) of conveyor haulage are around one-third that of a comparable TS system. However, when capital expenditure (CAPEX) is considered, the reduction in the total cost of operation over the mine life is around 50 per cent [14].

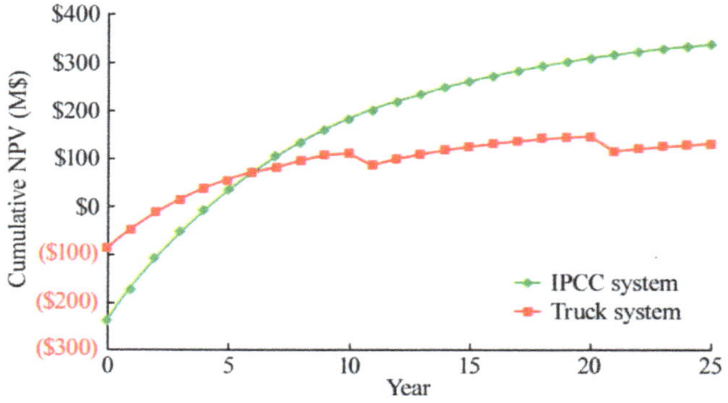

Figure 8. The cumulative net present value comparison of TS and IPCC systems [11].

In the meantime, there are several disadvantages to IPCC versus TS:

1. Flexibility

Its flexibility is the most significant factor that hinders the commercial take-up of IPCC systems when a mine considers an available mining system. Of the four types of IPCC systems, SM-IPCC shows the best flexibility, while FM-IPCC is the worst. Mine design, relocation and capability are limiting IPCC application.

(a) Mine design limitation. The decision-makers must cater to the installation requirements of the IPCC systems when they design the mine layout. Take FM-IPCC as an example, the optimization of ultimate pit limit (UPL), considering the geometric constraints connected with the installation of FM-IPCC systems, is one study field that requires substantial further investigation [11]. Throughout each sinking phase of a mine, truck haulage may still be required, but the distance of the haul may be decreased by deploying and scheduling the trucks to dump into the fully mobile crusher close to the mining activity [14].

(b) Relocation limitations. The IPCC has its specific extraction sequence. It is crucial to design its optimal location and relocation strategy to minimize operating costs. Mine designers need to trade-off large bench widths against production for an optimal location and relocation strategy [37]. For instance, because FM-IPCC systems are better suited to flat or gently dipping applications such as coal overburden or iron ore mining, it reduces the ability of a mine to switch mining to a different zone to adapt to unforeseen changes in market conditions or geology [14].

(c) Capacity limitations. Compared with the TS system, IPCC systems cannot be scaled up or down as mining requirements change [38]. This is because IPCC's major components (crusher, conveyor, spreader/stacker) have their own capacity limitations. An IPCC system also has a rated capacity, which reduces the ability to scale mining rates up or down according to market conditions.

2. Reliability

As IPCC systems are a series of connected systems, an unplanned delay or maintenance outage in one piece of equipment will affect the throughput of the entire system [14]. The availability of the whole IPCC system depends not only on the availability of the crusher but on the availability of each of the conveyors that comprise the whole system; the more components there are, the lower the reliability of IPCC systems [24].

3. Material requirements

The material requirements of IPCC transportation focus on material size and material properties. In order to transport material via conveyor, particle size distribution should be such that the largest material does not exceed approximately one-third of the belt width [14]. On the material properties side, the ability to sustain high throughput rates (4000–10,000 t/h) through a mobile crusher is key to IPCC system performance. Comprehensive knowledge of the material characteristics of the deposit and waste rock is required to specify the correct crusher type [14].

4. Contractual constraints

IPCC systems are not available as off-the-shelf solutions. The current approach for acquiring IPCC systems is via engineering, procurement, construction and management contract, which adds cost and delay to a mining project. This procedure will likely change once IPCC technology matures and gains greater acceptance [14].

Overall, the comparison results indicated that the IPCC system is superior for mining activities requiring strict environmental management, long lifespan, high production rate, and long-haul distances [17]. Generally speaking, the use of an IPCC continuous mining system will lower the energy consumption and significant emissions in the haulage sector of a mine, as well as reduce the cost of the haulage mine sector as a whole by millions of dollars, which will ultimately boost the mining sustainability and economy [13].

5.2. Trolley Assist Systems

5.2.1. Theory of Trolley Assist

After the oil supply crisis in the mid-1970s, the surface mining industry turned its attention to this fuel-saving technology. Several surface mines equipped with large off-highway electric trucks considered introducing TA into their operations. The overall view of the TA system is shown in Figure 9. As a solution that is a practical first step on the path to low-emission mine sites, TA is a proven technology capable of providing external electrical power to diesel-electric equipment. Recent advances in electric control technology have made this type of haulage an attractive alternative to conventional diesel-electric haulage [43].

The objective of mine decision-makers is to transport the highest volume of payload per hour while minimizing operating costs over one haul cycle of the trucks within acceptable risk boundaries [44]. Therefore, the power supply module which produces power from a diesel engine may be integrated with overhead trolley electricity to achieve further fuel savings [45,46]. The TA system is the most cost-effective on the ramps, where the majority of the total energy is used [44].

As Figure 10 shows, after operators manoeuvre diesel-electric trucks leaving the workface to arrive at the trolley ramp, operators determine the most appropriate time and approach speed to enter trolley mode to raise the pantograph. The truck switches to trolley electricity when the pantograph is activated and connected to the overhead power lines. Additionally, the truck's diesel engine enters idle mode, significantly saving fuel energy and reducing CO_2 emissions [45]. Because the electric wheel motor power commonly exceeds engine power, the electric wheel motors' full power capacity can achieve accelerating speed in trolley mode [47]. From a power aspect, with pantographs, diesel-electric hauling trucks could draw power from an overhead trolley line. However, diesel-electric power is still required in the pit, surrounding the loader/crusher, during hauling level segments, and on return travel [44].

(**a**)

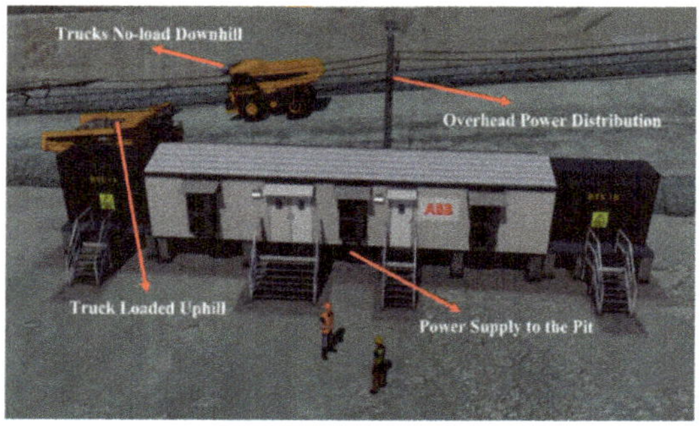

(**b**)

Figure 9. (**a**)The overall view of Trolley Assist system (source: ABB). (**b**) The detail of Trolley Assist on the ramp.

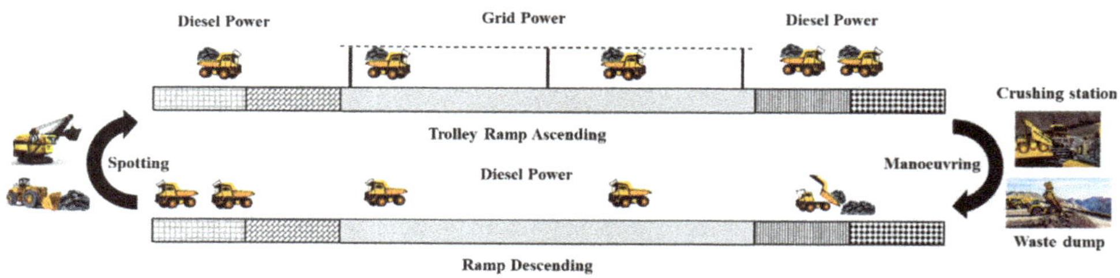

Figure 10. A schematic of a typical TA operation and power process.

5.2.2. Configuration of Trolley Assist

Trolley Assist systems supplement the power requirements of diesel-electric haul trucks via an external power source. Diesel-electric haul trucks are powered by a diesel engine generating an Alternative Current (AC) that powers the rear wheel motors to deliver torque to the wheels. Under Trolley Assist, the wheel motors are powered by an external Direct Current (DC) power source, commonly an overhead power distribution system. TA systems consist of three subsystems: power supply to the pit, overhead power distribution, and trucks with TA capability (Figure 11) [48].

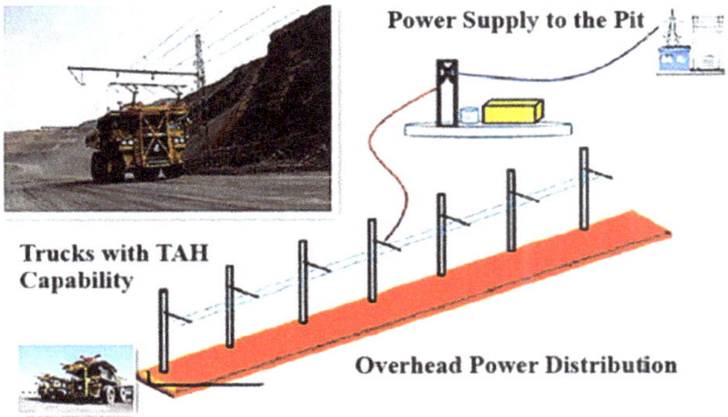

Figure 11. System design of TA.

1. Power Supply to the Pit

Transmission lines from a utility source deliver AC power to strategically positioned rectifier substations, providing DC power to the trolley line along the haul road. Rectifier substations should be skid-mounted for mobility to accommodate changes in the haul route [48]. The rectifier substations deliver power to the overhead power distribution system along the haul route.

2. Overhead Power Distribution

Overhead power distribution is achieved by a catenary system that supplies electrical power along the haul road. The catenary system allows the trucks to drive underneath and connect to the DC power. The voltage supplied by the catenary system depends on the wheel motors in the trucks using TA [49]. The catenary system is supported by poles spaced approximately 20–30 m along the haul road. The supporting poles' actual design and spacing depend on the haul road characteristics.

3. Trucks with Trolley Assist Capability

Figure 12 illustrates a truck operating under TA. The truck conversion was required in trolley mode when Trolley Assist systems were implemented, which in some cases required rewiring the wheel motor circuit to operate in series during TA. A pantograph system was required to connect the truck's electrical circuit to the overhead power, which operators determine the most appropriate times to raise or lower.

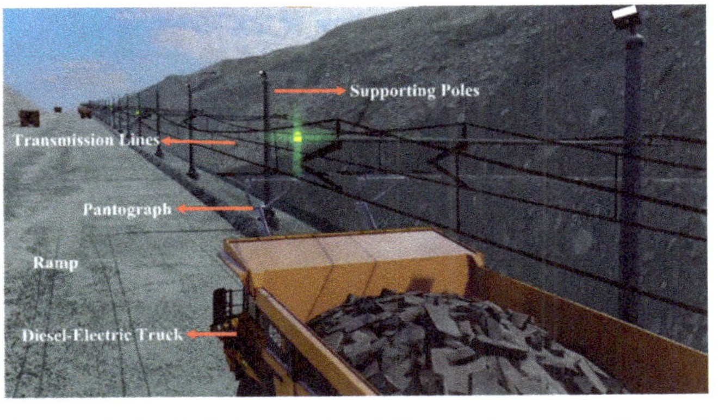

Figure 12. Truck with Trolley Assist capability operating on ramp (Source: ABB).

5.2.3. Advantages and Disadvantages of Trolley Assist System

In each of the TA applications, there was widespread publicity of various benefits resulting from the conversion to TA. The performance estimates of the TA system were derived from an analysis of several South African fleets that had used the technology in the past [48]. According to available reports and mine site production experience, TA systems offer the following benefits compared to TS alternatives. The advantages of TA are:

1. Reduced Emissions

The most crucial advantage is significantly reducing truck fleet CO_2 emissions [49]. According to ABB statistics, a single mining truck emits around 1000 tons of CO_2 per year [28]. Trolley Assist systems are most effective on the ramps, where most diesel energy is consumed, and the emissions are emitted.

2. Reduced diesel fuel consumption

A substantial saving in diesel fuel is made possible on the ramp with the use of a TA. According to a study case, the usage of TA decreases diesel consumption for ramp haulage by 19 litres per kilometre per truck [50]. The value of the resulting savings depends on the relative pricing of diesel fuel versus electricity. Indeed, relatively inexpensive electricity is more popular than diesel, and fossil fuel prices have continued to rise in the past few years. On the other hand, it is interesting to note that the several mines converted to TA systems are located in southern Africa, where the region faced the most severe oil supply problems during the 1970s. Reduced diesel fuel consumption can relieve the pressure on remote mines with oil supply problems [51].

3. Productivity improvements

Electric motors also offer more torque at low speeds than traditional combustion engines [52], which means electrified trucks are able to accelerate faster and provide better speed performance on a ramp [49]. Due to the trucks' higher gradient capability while operating under grid power, the TA system makes it simpler to access the deeper portions of the mine as mining depth increases. The use of TA on uphill hauls usually increases haulage system productivity. The report shows the resulting benefits are truck speed increases on a ramp from 13 km/h to 27 km/h [48]. The overall increase in truck productivity depends on the relative length and road grade of the uphill ramps in the haul cycle, and this productivity increase favours the use of TA for long uphill hauls.

4. Increase engine and wheel motor life

All the mines with TA systems have reported a substantial increase in engine and wheel motor life cycles and running time. Increased speed on ramps results in shorter times during which the wheel motors are at the full load; thus, motor overheating is less likely to occur. With the same motors, longer ramps may be negotiated without undue wheel motor wear, thus improving the haulage system for deep pit capacity.

5. Reduced fleet size

In a TA system, a single truck can complete one cycle in a shorter time due to its higher speed, which means the TA fleet can transport the higher possible payload per hour. Therefore, reducing the number of required trucks is likely to achieve production requirements. Fewer trucks lead to reduced maintenance and workforce costs, plus reduced capital expenditure.

6. Lower maintenance cost

Under TA haulage, a truck's diesel engine idling reduces the duty on the engine and increases the engine life. From a single truck perspective, trolley trucks need less maintenance than conventional diesel trucks because of reduced truck engine maintenance and fewer overhauls. On the downside, it is considered that the savings would be offset by the increased cost of electric wheel maintenance and trolley system maintenance [50].

7. Lower overall operating cost

Due to infrastructure and truck retrofit costs, although the TA system requires high upfront capital expenditure, the overall operating cost can be lower than conventional diesel trucks because TA is capable of reducing emissions (depending on carbon tax policy), energy consumption, the number of trucks, and maintenance costs.

In the meantime, there are several disadvantages to TA versus T:

1. High upfront capital outlay

The TA system is more complex than the conventional TS system with respect to the infrastructure of power supply to the pit, overhead power distribution, and the retrofit of trucks with TA capability, which means a high upfront capital outlay. According to research, the infrastructure cost per truck for adaptive measures and TA systems, which need an overhead cable, is around 75% of the overall truck price [53].

2. Mine design and planning restriction

The most significant advantages of off-highway haulage are its flexibility to mine schedule and ease of adjustment to a specific feature of the mined deposit. The installation of a TA system has imposed several restrictions on the flexibility of mine design and planning. While the trolley lines can be relocated, the relocation requires a skilled workforce, specialized equipment, and time. Time lost during critical stages of mining operations may have severe financial consequences. The cost and restrictions associated with the trolley shifting operation are likely to discourage frequent haul road relocations and restrict traffic patterns. Trolley shifting is another challenge in trolley ramp structural design, which affects trolley system performance. While not all the restrictions may apply to each mining situation, the associated costs for those that do apply should be evaluated and accounted for. More importantly, the TA system still preserves the majority of benefits in flexibility by using diesel-electric trucks. Although TA system flexibility is less than the conventional TS system, it is higher than IPCC systems.

3. Trolley Assist system maintenance

The reported experience with the existing TA system indicates that the distance between the trolley wire and the haul road surface must be closely controlled [54]. The need to maintain smooth haul routes and tight tolerances between the haul road surface and overhead lines is currently experienced with trucks operating under a TA system where wheel path wander is minimal due to the requirement to position the pantograph under the power lines. In this case, more ancillary equipment is necessary to maintain road quality. In addition, the maintenance costs include infrastructure maintenance and inventory of related spare parts and materials.

4. System capacity

The trolley sections have a limited capacity and are only able to accommodate a given number of trucks due to grid power limitations. When this number is on the section, the next truck cannot be accepted, and it must wait or travel powered by its diesel engine. Furthermore, slow-moving equipment, such as water trucks and graders, may slow the Trolley Assisted truck down. Truck schedules need to balance production tasks and maintenance requirements while considering TA technology's capacity. Bunching of trucks frequently occurs in TA operations, resulting in potential productivity loss. Therefore, considering system capacity limitations, the TA fleet needs a more effective dispatch strategy.

5. Access to Electricity

Installation of a TA system will require additional electrical power capacity. The TA system becomes an option for mine sites that can increase their electrical power capacity (i.e., readily available power or excess capacity). The adoption of the truck haulage system outside South Africa indicates that TA may be economically feasible in situations without very high diesel-to-electricity cost ratios. For remote mines, renewable energy sources, such as wind turbines and solar PV, may be used as alternatives to fulfil the electrical power requirement of the TA system, which will be driven by decarbonization.

6. Operator requirement

Operator training is essential to the truck haulage system because operators determine the most appropriate times to raise or lower the pantograph. The higher truck speeds combined with narrower steer paths demand more excellent skill and concentration from an operator. Greater awareness of the truck's dimensions is required to avoid collision with the catenary system supporting poles. If the truck loses contact with the trolley wire on the ramp due to erroneous driving, it will cause a severe bunching phenomenon because of lower speed and loss of potential productivity.

In most cases, a permanent, long-haul route with TA on the ramp out of the pit will result in the best economic benefit for Trolley Assist. The TA system's economic feasibility depends on several factors, including the availability of alternative electricity, diesel fuel and electricity costs, the cost of employing operators, and resulting maintenance requirements.

5.3. *Battery Trolley Systems*

5.3.1. Theory of Battery Trolley

The mining industry is working on a series of projects to achieve zero-emissions fleet requirements. Battery Trolley deployment is one such option [4,46]. Battery Trolley aims to offer a haulage mining system using the full source of electrical power as a decarbonization technology through autonomous high-intensity battery-electric trucks, TA systems and energy recovery systems.

5.3.2. Technology Uptake

It is advanced technology development that gives BT a chance to be a reality. Battery-electric power, autonomous deployment, TA and energy recovery technologies are the critical drivers for the BT to achieve the decarbonization pathway, which are core components in the future plans for deeper phases [31].

1. Battery-electric power technology

Electromobility, defined as the development and usage of electric-powered vehicles, is an industry-wide technical trend [31]. BEVs are one of the choices available to accomplish ambitious decarbonization goals. New battery designs with superior usage performance and lower cost will boost BEVs' competitiveness in the mining sector. Battery electric trucks have fewer mechanical systems and control logic than conventional hybrid ones, which results in reduced failure rates and more straightforward maintenance [45]. Nevertheless, battery size, energy density, battery swapping and charging, battery health and management are challenges facing the mining sector when thinking about applying battery-powered trucks.

2. Autonomous technology

According to statistics collected by GlobalData, by May 2022, there were 1068 autonomous haul trucks operating worldwide, a 39% yearly growth. Caterpillar and Komatsu supply 86.5% of the trucks monitored by the Mining Intelligence Centre, with the 793F and 930E being the two OEMs' most popular models, respectively [55]. That is because autonomous solutions can improve safety, equipment availability, and overall productivity on any mine site without machine operators sitting in the cab. As for the BT, determining the most appropriate times by leveraging autonomous technology to raise or lower the pantograph is the best option. BT systems are capable of taking advantage of autonomous trucks from both safety and productivity perspectives.

3. Trolley Assist technology

Battery electric vehicles are one option for mining trucks. However, in order to overcome battery size and energy density defects, mining trucks need TA technology to provide ascending energy on an uphill where the most energy is consumed. TA technology makes BT available by offering electric power to battery trucks, which enables battery trucks to haul for a long time.

4. Energy recovery system

The BT is able to leverage an energy recovery system to recuperate braking energy, which is used to charge the onboard battery when returning downhill [47]. The depth alterations connected with mining development bring significant variations in haul cycles and recoverable potential energy per cycle [56].

5.3.3. Battery Trolley Advantages and Disadvantages

Battery Trolley makes it possible to achieve the first zero-emissions truck fleet as a green solution, which is available to remove the reliance on fossil fuels by using battery-electric power in mining haulage systems. Except for decarbonization, reducing energy costs and TA's advantages, the BT can achieve lower maintenance costs for a single truck without a diesel engine. Additionally, from the overall mine operating life, the operating costs of BT are less than the conventional diesel truck fleet because of using electricity as end-use energy, which is similar to IPCC.

In spite of the advantages associated with BT, decision-makers may be reluctant to use it for some reasons. From diesel-electric to battery-electric power, this transition would significantly increase the mine's electricity cost and demand, as well as the power infrastructure and station capital expenditure. Additionally, the battery truck fleet has to face many challenges, such as battery size and performance, high upfront capital outlay, feasibility, availability, capability, truck fleet dispatching, mine design restrictions, and ancillary equipment maintenance schedule arrangement.

5.3.4. Battery Trolley Systems Configurations

Like IPCC systems, there are three possible configurations for BT. Each type has its pros and cons, which can be used in unique mining situations.

1. Dynamic charging BT configuration

Dynamic charging technology enables the ability for grid power to be used to power the electric drive motors and charge the onboard vehicle battery simultaneously. The dynamic charging BT consists of the battery-electric truck, the TA systems and dynamic charging technology.

Figures 13 and 14 are, respectively, the dynamic charging BT systems operational process and power source. Battery-electric trucks load and haul with battery power, switching to trolley mode after arriving at the trolley ramp. The battery consumes energy at a much lower rate for cooling and idling. At the same time, the grid power is simultaneously used to charge the on-board battery and provide the wheel motors output power on the trolley ramp. When the battery-electric truck comes onto an ex-pit flat road, it returns to battery power mode to complete hauling, queueing, dumping and returning manoeuvres. The battery-electric truck then enters energy recovery mode on the downhill ramp. The energy recovery system transforms truck braking power into electric energy that can be stored on the battery. The battery-electric truck then reuses battery power to return to the loading point.

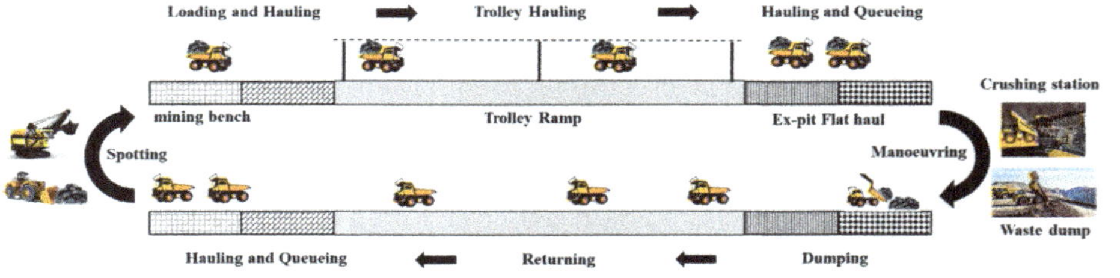

Figure 13. A schematic of dynamic charging BT systems operational process.

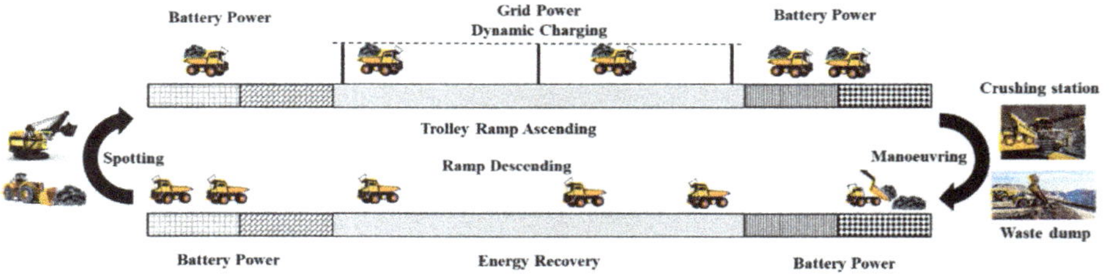

Figure 14. A schematic of dynamic charging BT systems power source.

2. Stationary charging BT configuration

In stationary charging method, a battery station is necessary for battery charging/swapping. As for choosing charging method or swapping method, it depends on charging C-rate and swapping time. The location of battery station is selected on the crest of pit for providing enough permanent room to build infrastructure and park trucks. The stationary charging Battery Trolley consists of the battery-electric truck, the TA systems and battery station.

Figures 15 and 16 are, respectively the stationary charging BT systems operational process and power source. Battery-electric trucks load and haul with battery power, switching to trolley mode after arriving at the trolley ramp. The battery consumes energy at a much lower rate for cooling and idling. At the same time, the grid power is capable of providing the max wheel motors output power to operate in a faster speed on the trolley ramp. When the battery-electric truck comes onto an ex-pit flat road, it returns to battery power mode to complete hauling, queueing, dumping and returning manoeuvres. According to on-board battery size design and energy consumption, the battery-electric truck needs to charging/swapping battery within each cycle or every two/three cycles. The battery-electric truck then enters energy recovery mode on the downhill ramp. The energy recovery system transforms truck braking power into electric energy that can be stored on the battery. The battery-electric truck then reuses battery power to return to the loading point.

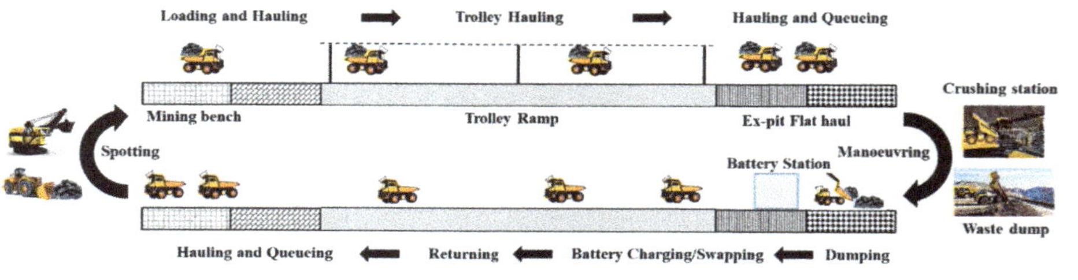

Figure 15. A schematic of stationary charging BT systems operational process.

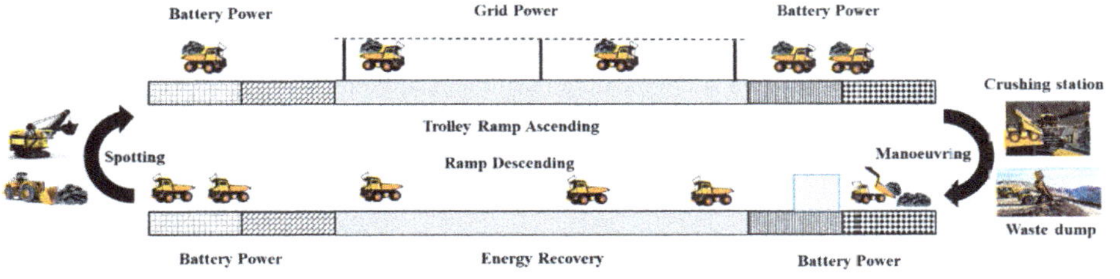

Figure 16. A schematic of a typical BT systems power source.

3. Dual trolley BT configuration

Research shows that for downhill hauls, a bidirectional substation enables energy feedback to the grid [44]. It is reasonable to install a dual trolley system for better energy capture performance in a BT system: the uphill ramp trolley enables the ability for grid power to be used to power the electric drive motors while the downhill ramp trolley captures braking energy and returns it to the grid. The dual trolley BT consists of battery-electric trucks and a double trolley system.

Figures 17 and 18 are, respectively, the dual trolley BT systems operational process and power source. Battery-electric trucks load and haul with battery power, switching to trolley mode after arriving at the trolley ramp. The battery consumes energy at a much lower rate for cooling and idling. At the same time, the grid power is capable of providing the max wheel motors output power to operate in a faster speed on the trolley ramp. When the battery-electric truck comes onto an ex-pit flat road, it returns to battery power mode to complete hauling, queueing and dumping manoeuvres. According to on-board battery size design and energy consumption, the battery-electric truck needs to charging/swapping battery within each cycle or every two/three cycles. The battery-electric truck needs to charging/swapping batteries on returning travel when it passes a battery station located on the pit's crest. The battery-electric truck then enters energy recovery mode downhill by engaging the trolley line, which captures braking energy back to the grid. The battery-electric truck then reuses battery power to return to the loading point.

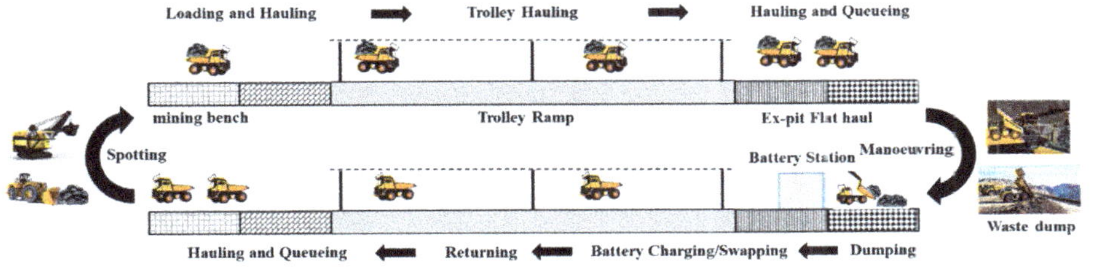

Figure 17. A schematic of dual trolley BT systems operational process.

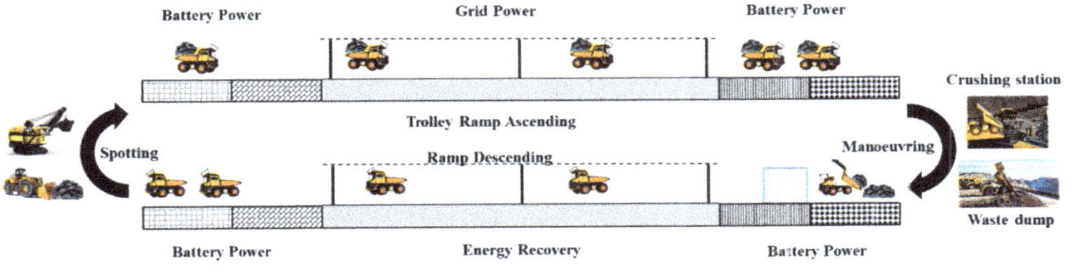

Figure 18. A schematic of dual trolley BT systems power source.

6. Discussions

In order to achieve optimum make-decision in mining haulage systems, it is necessary to use the mining system analysis method for evaluating each mining system parameter in Table 2.

Table 2. Comparison between diesel TS, IPCC, TA and BT.

Parameter	Diesel TS	SF/M IPCC	FM-IPCC	TA	Dynamic Charging BT	Stationary Charging BT	Dual Trolley BT
Flexibility	High	Medium	Low	Medium	Low	Medium	Medium
Energy Efficiency	Low	Medium	High	Medium	High	High	High
CAPEX	Low	High	High	High	High	High	High
OPEX	High	Medium	Low	Low	Low	Low	Low
Maintenance Requirements	High	Medium	Low	Medium	Medium	Medium	High
Service Life	Short	Medium	Long	Long	Long	Long	Long
Additional Infrastructure	No	No	No	Yes	Yes	Yes	Yes
Refuelling/ Recharging/ Swapping	Fast	None	None	Fast	None	Low	Low
Emissions	High	Low	None	Low	None	None	None
Heat Generation	High	Medium	Low	Medium	Low	Low	Low
Environmental Footprint (Noise/Dust/ DPM/Vibration)	High	Medium	Low	Medium	Low	Low	Low
Reliability	High	Medium	Low	Medium	Low	Medium	Low
Scalability	High	Low	Low	Medium	Low	Medium	Low
Capability	No	Yes	Yes	Yes	Yes	Yes	Yes
Safety	Low	Low	Medium	Low	Medium	Medium	Medium

According to Table 2, diesel TS shows the best performances in flexibility, CAPEX, refuelling, reliability, scalability, and capability, which explains why classic TS are prevalent in all kinds of greenfield and brownfield mining projects. IPCC is capable of mitigating the TS disadvantages from energy efficiency, maintenance, refuelling, emissions, heat generation, and environmental footprint points. However, flexibility, CAPEX, reliability, scalability, and capability characteristics are the constraints for IPCC, especially FM-IPCC, to large-scale applications in mine sites. Due to diesel-electric power and trolley limitations, TA shows medium performance in almost all parameters. In the dynamic charging alternative, because the onboard battery energy source is from grid charging uphill and energy capture downhill, the battery-electric trucks cannot complete one haul cycle without enough trolley lines charging. Therefore, dynamic charging BT has lower flexibility, reliability, scalability and capability compared with stationary charging BT, while no recharging/swapping battery need in the battery station is the most significant merit for dynamic charging BT systems. Because of flexibility limitations and considerable capital outlays, dual trolley BT is unlikely to be popular in large-scale BT deployment. However, dual trolley BT is suitable for some unique mine site conditions like super-depth copper mines.

7. Conclusions

The mining industry is now at a crossroads with surface mining fleets as it works to meet interim reduced emissions and final net-zero targets. A big part of that is moving away from diesel to electricity alternatives. This paper depicts the various haulage systems from diesel-based power trucks to electric-based power IPCC, diesel-electric power TA systems and battery-electric power BT systems. IPCC and TA are ramping up due to reasonable economic and emission reduction, whilst trucks operating in conjunction with a conceptual BT system could decarbonise haulage mining systems in open pit mines. All these haulage systems are interrelated and complementary. They cannot be determined in isolation, which requires further comparison and analysis of their mine sites' practice performance, whereby all advantages and disadvantages are considered simultaneously. Large open pit mines may require a combination of different systems, e.g., SM-IPCC and BT systems, to achieve the decarbonization haulage system.

Data Availability Statement: Not Applicable.

Conflicts of Interest: The authors declare no conflict of interest.

Abbreviations

TS	Truck-Shovel
IPCC	In-pit Crushing and Conveying
TA	Trolley Assist
BT	Battery Trolley
ERSs	Energy Recovery Systems
CAPEX	Capital Expenditures
OPEX	Operating Expenses
STEPS	Stated Policies Scenario
APC	Announced Pledges Case
SDS	Sustainable Development Scenario
NZE	Net Zero Emissions
IEA	International Energy Agency
GHG	Greenhouse Gas
PV	Photovoltaic
NEM	National Electricity Market
TGP	Terminal Gate Price
DPM	Diesel Particulate Matter
AHTs	Autonomous Haulage Trucks
FIPCC	Fixed In-pit Crushing and Conveying
SFIPCC	Semi-Fixed In-pit Crushing and Conveying
SMIPCC	Semi-Mobile In-pit Crushing and Conveying
FMIPCC	Fully Mobile In-pit Crushing and Conveying
UPL	Ultimate Pit Limit
AC	Alternative Current
DC	Direct Current
BEVs	Battery-electric vehicles

References

1. Plazas-Niño, F.A.; Ortiz-Pimiento, N.R.; Montes-Páez, E.G. National energy system optimization modelling for decarbonization pathways analysis: A systematic literature review. *Renew. Sustain. Energy Rev.* **2022**, *162*, 112406. [CrossRef]
2. International Energy Agency (IEA). Net Zero by 2050—A Roadmap for the Global Energy Sector. Available online: https://iea.blob.core.windows.net/assets/deebef5d-0c34-4539-9d0c%2010b13d840027/NetZeroby2050ARoadmapfortheGlobalEnergySector_CORR.pdf (accessed on 1 November 2022).
3. International Energy Agency (IEA). World Energy Outlook. 2021. Available online: https://www.iea.org/reports/world-energy-outlook-2021 (accessed on 1 November 2022).
4. Zuliani, J.E.; Guilbaud, J.; Carreau, M. Decarbonization of remote mine electricity supply and vehicle fleets. *CIM J.* **2021**, *12*, 135–148. [CrossRef]
5. Department of Industry, Science, Energy and Resources. Australia's Emissions Projections. 2020. Available online: https://www.industry.gov.au/data-and-publications/australias-emissions%20projections-2020 (accessed on 1 November 2022).
6. Vatalis, K.I.; Avlogiaris, G.; Tsalis, T.A. Just transition pathways of energy decarbonization under the global environmental changes. *J. Environ. Manag.* **2022**, *309*, 114713. [CrossRef] [PubMed]
7. Burke, P.J.; Beck, F.J.; Aisbett, E.; Baldwin, K.G.; Stocks, M.; Pye, J.; Venkataraman, M.; Hunt, J.; Bai, X. Contributing to regional decarbonization: Australia's potential to supply zero-carbon commodities to the Asia-Pacific. *Energy* **2022**, *248*, 123563. [CrossRef]
8. Australian Renewable Energy Agency. Renewable Energy in the Australian Mining Sector White Paper. Available online: https://arena.gov.au/assets/2017/11/renewable-energy-in-the-australian-mining-sector.pdf (accessed on 1 November 2022).
9. Department of Industry, Science, Energy and Resources. Australian Energy Update. 2021. Available online: https://www.energy.gov.au/publications/australian-energy-update-2021 (accessed on 1 November 2022).
10. Parra, A.; Morales, N. Qualitative Review of Alternative Transport Systems for Open Pit Mines. Available online: https://delphoslab.cl/Publicaciones/2016/PM2_M2016.pdf (accessed on 1 November 2022).
11. Nehring, M.; Knights, P.F.; Kizil, M.S.; Hay, E. A comparison of strategic mine planning approaches for in-pit crushing and conveying, and truck/shovel systems. *Int. J. Min. Sci. Technol.* **2018**, *28*, 205–214. [CrossRef]
12. Calvo, G.; Mudd, G.; Valero, A.; Valero, A. Decreasing ore grades in global metallic mining: A theoretical issue or a global reality? *Resources* **2016**, *5*, 36. [CrossRef]

13. Purhamadani, E.; Bagherpour, R.; Tudeshki, H. Energy consumption in open-pit mining operations relying on reduced energy consumption for haulage using in-pit crusher systems. *J. Clean. Prod.* **2021**, *291*, 125228. [CrossRef]
14. Dean, M.; Knights, P.; Kizil, M.S.; Nehring, M. Selection and planning of fully mobile in-pit crusher and conveyor systems for deep open pit metalliferous applications. In Proceedings of the Third International Future Mining Conference, Sydney, Australia, 4–6 November 2015; pp. 219–225.
15. Igogo, T.; Awuah-Offei, K.; Newman, A.; Lowder, T.; Engel-Cox, J. Integrating renewable energy into mining operations: Opportunities, challenges, and enabling approaches. *Appl. Energy* **2021**, *300*, 117375. [CrossRef]
16. Lezak, S.; Cannon, C.; Blank, K. *Low-Carbon Metals for a Low-Carbon World: A New Energy Paradigm for Mines*; Rocky Mountain Institute: Basalt, CO, USA, 2019.
17. Osanloo, M.; Paricheh, M. In-pit crushing and conveying technology in open-pit mining operations: A literature review and research agenda. *Int. J. Min. Reclam. Environ.* **2020**, *34*, 430–457. [CrossRef]
18. Soofastaei, A.; Karimpour, E.; Knights, P.; Kizil, M. Energy-Efficient Loading and Hauling Operations. In *Energy Efficiency in the Minerals Industry. Green Energy and Technology*; Awuah-Offei, K., Ed.; Springer: Cham, Switzerland, 2018. [CrossRef]
19. Aguayo, I.A.O.; Nehring, M.; Ullah, W. Simulating the productivity improvements of incorporating a surge loader into the open pit loading and haulage system. *Int. J. Min. Reclam. Environ.* **2022**, *36*, 1–18. [CrossRef]
20. Bao, H.; Zhang, R. Study on optimization of coal truck flow in open-pit mine. *Adv. Civ. Eng.* **2020**, *2020*, 8848140. [CrossRef]
21. Bakhtavar, E.; Mahmoudi, H. Development of a scenario-based robust model for the optimal truck-shovel allocation in open-pit mining. *Comput. Oper. Res.* **2020**, *115*, 104539. [CrossRef]
22. Wang, Q.; Zhang, R.; Lv, S.; Wang, Y. Open-pit mine truck fuel consumption pattern and application based on multi-dimensional features and XGBoost. *Sustain. Energy Technol. Assess.* **2021**, *43*, 100977. [CrossRef]
23. De Lemo Pires, D. Surface mining technology: Managing the paradigm shift. *Min. Eng.* **2013**, *6*, 36–41.
24. Hay, E.; Nehring, M.; Knights, P.; Kizil, M. Ultimate pit limit determination for semi mobile in-pit crushing and conveying system: A case study. *Int. J. Min. Reclam. Environ.* **2020**, *34*, 498–518. [CrossRef]
25. Moniri-Morad, A.; Pourgol-Mohammad, M.; Aghababaei, H.; Sattarvand, J. A methodology for truck allocation problems considering dynamic circumstances in open pit mines, case study of the sungun copper mine. *Rud.-Geološko-Naft. Zb.* **2019**, *34*, 4. [CrossRef]
26. Kumral, M.; Sari, Y.A. An application of possibilistic programming to production sequencing of mining parcels. *Int. J. Manag. Sci. Eng. Manag.* **2017**, *12*, 79–88. [CrossRef]
27. Abbaspour, H.; Drebenstedt, C. Environmental Comparison of Different Transportation Systems—Truck-Shovel and IPCCs—In Open-Pit Mines by System Dynamic Modeling. In *Proceedings of the 27th International Symposium on Mine Planning and Equipment Selection—MPES 2018;* ; Widzyk-Capehart, E., Hekmat, A., Singhal, R., Eds.; Springer: Cham, Switzerland, 2019. [CrossRef]
28. ABB-ABB Ability™eMine. For Your World, and Mine. In Proceedings of the Electric Mine Conference, Stockholm, Sweden, 14–17 September 2022.
29. Abbaspour, H.; Drebenstedt, C.; Dindarloo, S.R. Evaluation of safety and social indexes in the selection of transportation system alternatives (Truck-Shovel and IPCCs) in open pit mines. *Saf. Sci.* **2018**, *108*, 1–12. [CrossRef]
30. Bozorgebrahimi, E. The Evaluation of Haulage Truck Size Effects on Open Pit Mining. Ph.D. Thesis, University of British Columbia, Vancouver, BC, Canada, 2004.
31. Sánchez, F.; Hartlieb, P. Innovation in the mining industry: Technological trends and a case study of the challenges of disruptive innovation. Mining. *Metall. Explor.* **2020**, *37*, 1385–1399. [CrossRef]
32. Bellamy, D.; Pravica, L. Assessing the impact of driverless haul trucks in Australian surface mining. *Resour. Policy* **2011**, *36*, 149–158. [CrossRef]
33. Dzakpata, I.; Knights, P.; Kizil, M.S.; Nehring, M.; Aminossadati, S.M. Truck and shovel versus in-pit conveyor systems: A comparison of the valuable operating time. In Proceedings of the 16th Coal Operators' Conference, Mining Engineering, University of Wollongong, Dubai, United Arab Emirates, 10–12 February 2016; pp. 463–476.
34. Ritter, R.; Herzog, A.; Drebenstedt, C. Automated dozer concept aims to cut IPCC downtime. *Eng. Min. J.* **2014**, *215*, 52–55.
35. Terezopoulos, N. Continuous haulage and in-pit crushing in surface mining. *Min. Sci. Technol.* **1988**, *7*, 253–263. [CrossRef]
36. Abbaspour, H.; Drebenstedt, C.; Paricheh, M.; Ritter, R. Optimum location and relocation plan of semi-mobile in-pit crushing and conveying systems in open-pit mines by transportation problem. *Int. J. Min. Reclam. Environ.* **2019**, *33*, 297–317. [CrossRef]
37. McCarthy, M.; Cenisio, B. Update on studies for implementation of IPCC at the Moatize coal project. In Proceedings of the IPCC 2013, Cologne, Germany, 25–26 June 2013.
38. Czaplicki, J. *Shovel-Truck Systems: Modelling, Analysis and Calculations*; CRC Press/Balkema, Taylor and Francis Group: London, UK, 2009.
39. McCarthy, B. Evaluating the lack of flexibility against the benefits of in-pit crushing and conveying. In Proceedings of the Optimizing Mine Operations Conference, Toronto, ON, Canada, 10–13 September 2013.
40. Schroder, D.L. The use of in-pit crushing and conveying methods to significantly reduce transportation costs by truck. In Proceedings of the Coaltrans Asia Conference, Bali, Indonesia, 8–11 June 2003.
41. Raaz, V.; Mentges, U. In-pit crushing and conveying with fully mobile crushing plants in regard to energy efficiency and CO_2 reduction. In Proceedings of the IPCC 2011, Belo Horizonte, Brazil, 28 June 2011.
42. Foley, M. In-pit crushing: Wave of the future. *Aust. J. Min.* **2012**, 46–53.

43. Lake, D.M.; Brzezniak, W. Trolley assist truck haulage cuts fuel and increases productivity. Part two. *Aust. Coal Min.* **1981**, *3*, 11, 13, 15–16.
44. Mazumdar, J. Performance improvement of mining haul trucks operating on trolley systems. In Proceedings of the 2011 IEEE Industry Applications Society Annual Meeting, Orlando, FL, USA, 9–13 October 2011; pp. 1–6.
45. Moore, P. Surface Mining Trucks. *Int. Min.* **2022**, *17*, 50–67.
46. Gleeson, D. Battery and Electric. *Int. Min.* **2022**, *17*, 26–47.
47. Engineering and Mining Journal. *Freeport Taps Wärtsilä for 128-MW Plant for Grasberg: Engineering, Geology, Mineralogy, Metallurgy, Chemistry, etc.*; Engineering and Mining Journal: Jacksonville, FL, USA, 2020; Volume 221, pp. 48–49.
48. Crosson, C.C.; Sumner, H.B. Trolley assisted truck haulage. United States. *Eng. Min. J.* **1982**, *183*. Available online: https://www.osti.gov/biblio/6730531 (accessed on 30 December 2022).
49. Morton, J. Truck Innovations Show Shape of Things to Come. *Eng. Min. J.* **2021**, *222*, 24–30.
50. Freeman, S.; Golosinski, T.S. Economic feasibility of introducing trolly assisted haulage to an Australian open pit mine. *Int. J. Surf. Min. Reclam.* **1994**, *8*, 141–144. [CrossRef]
51. Lake, D.M.; Brzezniak, W. Truck haulage using overhead electrical power to conserve diesel fuel and improve haulage economics. United Kingdom. *Min. Technol.* **1982**, *64*, 157–164.
52. Marsden, O.W.; Marsden, J.O. Potential Pathways for Mining Operations to Transition to Renewable Energy—A Case Study. *Min. Metall. Explor.* **2021**, *38*, 1689–1699. [CrossRef]
53. Ertugrul, N.; Kani, A.P.; Davies, M.; Sbarbaro, D.; Morán, L. Status of Mine Electrification and Future Potentials. In Proceedings of the 2020 International Conference on Smart Grids and Energy Systems (SGES), Perth, Australia, 23–26 November 2020; pp. 151–156.
54. Vorster, B.J. Trolley assist aids haulage at Iscor's Shishen iron ore mine in South Africa. *Min. Eng.* **1986**.
55. Caterpillar, Products Services. Available online: https://www.cat.com/en_US/articles/customer-stories/mining/autonomous-haulage-making-mining-safer-and-more-productive-today.html (accessed on 1 November 2022).
56. Terblanche, P.J. Technical and economic potential of energy recovery and re-use on board surface mining haul trucks. 2019. Available online: https://core.ac.uk/reader/200258120 (accessed on 1 November 2022).

mining MDPI

Article

Inclusive Urban Mining: An Opportunity for Engineering Education

Sofia L. Schlezak *,† and Jaime E. Styer *,†

Humanitarian Engineering and Science MS Program, Department of Engineering, Design, and Society, Colorado School of Mines, 1318 Maple St. Bld #4, Golden, CO 80401, USA
* Correspondence: schlezaksofia@gmail.com or slschlezak@mines.edu (S.L.S.); jaime.styer22@gmail.com or jstyer@mines.edu (J.E.S.)
† These authors contributed equally to this work.

Abstract: With the understanding that the mining industry is an important and necessary part of the production chain, we argue that the future of mining must be sustainable and responsible when responding to the increasing material demands of the current and next generations. In this paper, we illustrate how concepts, such as inclusiveness and the circular economy, can come together in new forms of mining—what we call *inclusive urban mining*—that could be beneficial for not only the mining industry, but for the environmental and social justice efforts as well. Based on case studies in the construction and demolition waste and WEEE (or e-waste) sectors in Colombia and Argentina, we demonstrate that inclusive urban mining could present an opportunity to benefit society across multiple echelons, including empowering vulnerable communities and decreasing environmental degradation associated with extractive mining and improper waste management. Then, recognizing that most engineering curricula in this field do not include urban mining, especially from a community-based perspective, we show examples of the integration of this form of mining in engineering education in first-, third- and fourth-year design courses. We conclude by providing recommendations on how to make inclusive urban mining visible and relevant to engineering education.

Keywords: urban mining; circular economy; sustainable development; engineering education; humanitarian engineering; community development

Citation: Schlezak, S.L.; Styer, J.E. Inclusive Urban Mining: An Opportunity for Engineering Education. *Mining* **2023**, 3, 284–303. https://doi.org/10.3390/mining3020018

Academic Editors: Juan M. Menéndez-Aguado and Mostafa Benzaazoua

Received: 2 March 2023
Revised: 5 May 2023
Accepted: 17 May 2023
Published: 18 May 2023

1. Introduction

The first reference to urban mining is claimed to be in *The Economy of Cities* by the Urban Theorist Jane Jacobs in 1969 [1]. In her piece, the author described future cities as mines with huge, rich, and diverse raw materials [1,2]. However, the origins of this concept are still under discussion [3], and differences in definitions arise based on the contrasting ideologies and priorities of stakeholders yielding the term.

In general, an "urban mine" is understood as the urban accumulation of anthropogenic materials aboveground [2,4], and "urban mining" could be interpreted as the activity that converts "wastes" into resources [5]. Aldebei et al. [2] understand urban mining as a metaphor for describing the same activities of prospecting, exploration, development, and exploitation as traditional mining. For the purpose of this article, we will utilize the following definition proposed by Cossu and Williams, "Urban Mining extends landfill mining to the process of reclaiming compounds and elements from any kind of anthropogenic stocks, including buildings, infrastructure, industries, products (in and out of use), environmental media receiving anthropogenic emissions, etc." [6] (p. 1).

The term "urban mining" is assumed to be applicable to many kinds of waste [5]. However, this work is based on two specific streams, namely construction and demolition waste (C&DW) and waste of electrical and electronic equipment (widely known as WEEE or e-waste). These are two of the most relevant anthropogenic sources in terms of quantity and economic incentive [7]. E-waste mainly motivates research and practice because of its high concentration of rare earth minerals, and buildings and infrastructure waste are the largest anthropogenic stock worldwide. In other words, C&DW is "the largest urban mine" [2] (p. 6).

As a direct consequence of the population growth, urbanization, and excessive consumption that characterize the last century, the exploitation of natural resources and the generation of waste have increased radically [5]. Under these circumstances, the concept of urban mining of anthropogenic wastes has been introduced for almost four decades as an alternative to the conventional way of extracting raw materials, which is particularly important to decrease their depletion and lower the mining footprint [8]. For example, managing electrical and electronic equipment under the circular economy approach can reduce the use of raw materials to produce new devices by up to 80% [7].

Urban mining also serves as an approach to sustainable waste management in cities and can be a source of new job opportunities for young people and/or immigrants [7]. As an anthropologist studying informal e-waste management in Tanzania observed, "(...) recycling offers a skilled vocation, with a sense of stepped progression, secure revenue and entrance into a social support network that sustains and enhances local lives" [9] (p. 7). Therefore, urban mining also has the potential to become inclusive by contributing to the production of goods and offering services while simultaneously pursuing social objectives to enhance the quality of life of vulnerable communities.

Although it is promising as an economic, social, and environmental activity, urban mining still has its limitations. The recent efforts in legislation to promote urban mining that have been implemented in Europe and other regions, for example, the WEEE Directive (2012/19/EU) and its recent amendments that have become international models for e-waste management [10–12], are not enough to deal with the 82% of electrical and electronic equipment that is not treated in a sound environmental manner [7,12] and the 35% of C&DW that still ends up in landfills globally [13]. The causes of these figures include low recovery efficiency rates as a consequence of inefficient product design, lack of development and effective implementation of regulations and certifications to promote the use of reused materials, negative perceptions about second-hand materials and products, lack of awareness about the benefits of urban mining and the impacts of e-waste and C&DW, space scarcity in urban centers to store materials, high costs of best-quality recycling processes that make it difficult to afford for small and medium-sized enterprises (SMEs), and high competitiveness of landfilling associated with immature local markets and poor economic incentives for the circular economy [7,14].

In Latin America, despite advances in this field, much work still needs to be undertaken to improve the low e-waste recovery rate below 2% [12] and C&DW recovery rate below 10% [15]. As a result, a significant part of the waste with economic potential is abandoned in open spaces [2] or exported, resulting in lower efficiency of the waste management systems [16]. Furthermore, this situation limits green job opportunities in the region, especially for informal waste pickers and recyclers, who currently play a key role in the circular economy [17,18].

In the last decade, many international organizations have focused on this issue, and financial resources have been allocated in this field, e.g., the 2018–2022 UNIDO-GEF PREAL project for the Environmentally Sound Management of POPs in Waste of Electronic or Electrical Equipment [19]. However, projects and research are, in general, led by stakeholders interested in industrial ecology, waste management, environmental health, and the circular economy rather than academics and researchers from the mining sector with an interest in the potential of urban mining as an alternative economic activity of material extraction and social empowerment [18].

While the traditional training of mining, metallurgical, and materials engineers might not focus on urban mining [18], we agree that "shortly, our society is undergoing an accelerating transition from virgin mining of linear economy to urban mining of circular economy" [20] (p. 104), and we suggest these groups should be part of this transition. Johansson et al. [18] claim that, as was the case for traditional and deep-sea mining, the development of technology could make urban mining attractive as an economic activity. They also highlight that even if the nature of mining changes, as with any mining activity, engineers and researchers specialized in materials composition, collection, extraction, separation, and recovery are crucial to overcoming current technological challenges for implementation.

In light of the need for further research and initiatives, there is a potential for science and engineering education to contribute to these global challenges. Some scholars reported a constant declining interest in mining studies worldwide [21] and proposed a focus on sustainable development to generate new competencies and subjects and promote innovative solutions and technologies by emphasizing environmental and social aspects [21]. Literature has reported the positive impacts of incorporating non-traditional mining areas into traditional engineering programs, for example, the case of incorporating artisanal and small-scale gold mining (ASGM) into the curriculum of an engineering college in the US [22,23], an approach that was also recommended by organizations, such as USAID and UNITAR as a crucial step towards formalization of the activity [24,25]. In this context, urban mining could also be proposed as an alternative to attract more students into the mining sector.

2. Objectives

Understanding that the mining industry is an important and necessary part of the production chain that should be aligned with international environmental agreements and goals, e.g., the United Nations 2030 Agenda and the Sustainable Development Goals, the future of mining must be sustainable and responsible when responding to the increasing material demands of the current and next generations. With this in mind, in this paper, we illustrate how concepts, such as inclusiveness and the circular economy, can come together in new forms of mining—what we call inclusive urban mining—that could be beneficial not only in mining engineering curricula and the mining industry but also for environmental and social justice efforts aimed at empowering vulnerable groups in regions, such as Latin America.

Population movement from rural areas to urban centers has created increasing demands for employment, often for individuals with low levels of education and literacy [26–28]. In this context, cities play a role in offering stable, secure, formal, economically sufficient, and dignified green jobs, including in the waste management sector [7]. Taking this into account, we begin by demonstrating how, if recognized, inclusive urban mining could present an opportunity to benefit society across multiple echelons, including empowering vulnerable communities (Section 5.1) and decreasing environmental degradation associated with extractive mining and improper waste management (Section 5.2). To do so, we use our research experiences in the C&DW and e-waste sectors in Colombia and Argentina to show how present and future urban miners are or can be empowered to build livelihoods out of treating what is traditionally seen as waste and how their path could be extended to prospective miners. Then, recognizing that most engineering curricula in this field (e.g., mining, environmental, and materials engineering) do not include urban mining, especially from a community-based perspective (Section 6.1), we show examples of the integration of this form of mining in engineering education in first-, third- and fourth-year design courses (Section 6.2). Finally, we conclude by providing recommendations for how to make inclusive urban mining visible and relevant to engineering education in different institutional contexts.

Given the few works found in the literature that introduce these ideas and perspectives, we challenge the status quo by proposing a conversation on a new paradigm that completely changes how we understand and treat material stocks. In this regard, we understand that the novelty of this study should be highlighted.

3. Description of Study Sites

3.1. C&WD in Colombia

In the past decade, there has been a rise in research and literature on C&DW management issues [29], and several countries, such as Germany, Spain, and Belgium are adopting strategies to treat and handle this type of waste [30]. However, Latin America lags in this area, and some countries, such as Colombia, despite generating vast amounts of C&DW, have not made noteworthy progress in managing it [31]. An estimated 35% of C&DW is disposed of in landfills without further treatment [32]. In their article, Colorado et al. attempt to obtain the first quantified values of C&DW in Colombia [13]. However, information on the management of C&DW in Colombia is very scarce, and Colorado et al. concluded that no reliable data depicting the amount of C&DW generated annually in Colombia exists. Similarly, most countries in Latin America do not collect data on the generation and quantification of C&DW [13].

Nevertheless, within a thesis project from 2003, author García Botero detailed C&DW in Bogotá, Colombia and examined if the sustainable development needs of the city are being met [33]. He concluded that approximately 99% of C&DW in the context of Bogotá is "useable", and a majority of this C&DW is made up of concrete, asphalt, brick, blocks, sand, gravel, earth, and mud. Furthermore, Méndez-Fajardo's article argues that recycling and reusing C&DW can produce significant positive impacts for citizens; however, these potential values are often overlooked [34]. These positive impacts can be seen at multiple echelons, including the environmental, social, cultural, economic, and even political level.

To study how to promote and support inclusive C&DW management in Colombia for the empowerment of low-income communities, we worked with Community A, a small, low-income community located just outside Girardot in the department of Cundinamarca. Girardot is a popular vacation spot and houses many recreational activities due to its warmer, tropical climate and proximity to Bogotá, the capital of Colombia which is home to over seven million people.

Unfortunately, not much is officially known about Community A. Based on our estimates, about 200 families live in the community, many of which do not have access to sewage systems. This site was considered relevant for our study because of their desire to take part in the project as well as the occupational profile of the inhabitants. While many of the men in the community work in construction practices, most of the women work in the informal sector (selling products, such as soda, avocados, and arepas from their homes or on carts either in the city or on the surrounding roads) or are unemployed. Furthermore, despite their occupational status, all the women in Community A are caretakers in their homes as well, for their children, parents, pets, and households. The local knowledge and the gender-related disparity in terms of job opportunities made Community A an interesting sample to study inclusive urban mining. The members in the community we spoke with wanted to learn how to extract value from C&DW and how they could make it profitable for themselves and their families. Thus, the goals of this project were refined with the guidance of the community, to increase education about C&DW and find a way to make this effort profitable.

3.2. E-Waste in Argentina

Despite the fact that Argentina had the highest generation of e-waste in 2019 (328 kt) out of the 13 countries studied by Wagner et al. [35], the management of this waste stream in Argentina is considered to be at a nascent stage, and little is known about it. Recent reports developed by national authorities confirmed a data gap [36], and the lack of regulations reveal that electronic waste management is a pending issue in the country. However, there are communities whose income depends on these materials. Some sources estimate that in 2017, nationally there were 600 people working in the informal e-waste recycling sector, and the number grew to around 2000 workers in 2019 [37]. A different source indicated that there were 2800 workers in 2019 in only 14 municipalities in the province of Buenos Aires [37].

The province of Buenos Aires and the City of Buenos Aires were selected as study sites in Argentina because they agglomerate the largest population and contain the enterprises and cooperatives that process the highest amount of e-waste [35,37]. In order to address the topic of inclusive urban mining, four cooperatives were studied. Cooperatives are "autonomous associations of persons united voluntarily to meet their common economic, social and cultural needs and aspirations through a jointly owned and democratically controlled enterprise" [38]. Three of the cooperatives under study are exclusively dedicated to e-waste, and one is a cooperative dedicated to solid waste but brings together workers who individually recover e-waste materials. Additionally, we included one university extension program currently offering e-waste management services in the province. The names and specific details of these facilities are protected, so they cannot be easily recognized.

- Facility A (cooperative): It started in 2018 as a solid waste cooperative, and since 2022, its members have decided to explore e-waste processing as an additional source of income. The e-waste sector now has five workers and one coordinator. They are in the process of formalizing their activity in relation to e-waste management.
- Facility B (cooperative): It is a solid waste cooperative that started almost ten years ago. They have more than 150 members, and almost half of them individually recover e-waste material from the streets. This cooperative is interested in e-waste, but its members do not yet have experience with this waste stream.
- Facility C (cooperative): With almost 20 workers, this cooperative is one of the province's most advanced small social businesses. They have already obtained legal permission to manage e-waste, their main activity.
- Facility D (cooperative): It has over 20 years of experience in the business and employs more than 25 workers. The cooperative is recognized as a formal e-waste operator and treats almost 1,400 tons of waste per year.
- Facility E (University extension program): It began as an academic extension program and now is one of the few e-waste operators in the province. Since 2009, they have trained 1168 students and treated 217 tons of e-waste.

4. Methods

4.1. Methodological Framework and Ethical Considerations

As the local knowledge of communities is crucial for developing sustainable and just solutions [39], our research methods were participatory and we took a qualitative approach to understand the neglected knowledge, expertise, and values of the communities we worked with, as well as the complex systems that shape their lives. Qualitative methodologies, such as semi-structured interviews, participant observation, focus groups as well as workshops, were utilized to build this understanding. Figure 1 summarizes the logical items of this study.

Both projects were approved by the Colorado School of Mines Human Subjects Research Team and exempted from the Institutional Review Board (IRB) process requirements. The research in Colombia was developed in partnership with a Colombian university, Corporación Universitaria Minuto de Dios or Uniminuto. The groups specifically working on this project were the Parque Científico de Innovación Social (PCIS)/the Social Innovation Science Park, a research group led by Civil Engineering and Occupational Health and Safety Professors with social justice aims entitled Ingeniero a tu Barrio, the international studies group in Uniminuto-Girardot, as well as communication specialists including Professor Martha Liliana Herrera Gutiérrez, who was responsible for translation, facilitation, and communication. Together we worked directly with a low-income community in Colombia, Community A, to study how recycling C&DW, specifically concrete, could empower them. Approval for this research was also obtained from local Colombian authorities, including Uniminuto's PCIS as well as their research ethics committee. Additionally, the research in Colombia adheres to Uniminuto's Social Innovation Route Framework [40], a five-phase community engagement framework developed by PCIS.

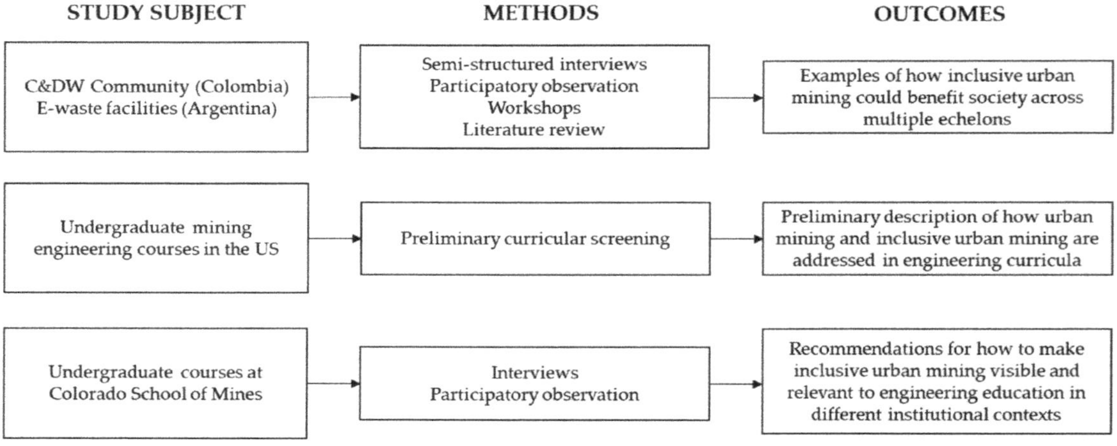

Figure 1. Schematic representation of the logical items of this study.

4.2. Semi-Structured Interviews

Throughout the six weeks of fieldwork on C&DW completed in Colombia during June and July 2022, we interviewed 17 women and 12 men from differing backgrounds, including low-income community members, community leaders, engineers, academics, students, waste management experts, and government officials. During each interview, translation services were provided by professors from Uniminuto. Within these interviews, we asked our interlocutors to describe their backgrounds, knowledge of concrete and C&DW, risks and barriers they believed could prevent recycling C&DW in low-income communities, and if any groups could be disproportionately affected by these risks. Additionally, as this project had specific social justice goals to contribute to women's empowerment in low-income communities, our focus was to speak mostly with women in the community to understand their interpretations of women's empowerment and define how they specifically wanted this project to empower them.

A series of exploratory interviews were conducted during June and July 2022 for the research on e-waste. A total of 15 government representatives, researchers, and members of e-waste cooperatives and programs were interviewed. The meetings, each lasting approximately one and a half hours, were held virtually. The interviews aimed to obtain preliminary information on the current situation of e-waste management in Argentina, with a particular focus on the province and the city of Buenos Aires. Participants were asked to describe their workplaces, e-waste-related practices and dynamics, the challenges the sector faces, and their opinions on past and current waste management strategies. Additional unrecorded interviews with e-waste workers and government officials were also conducted during the participatory observation visits described in the next section.

4.3. Participatory Observation

Upon beginning our fieldwork session in Colombia, the Uniminuto team immediately began facilitating meetings with Community A. As the Uniminuto team had already collaborated with this community in the past, many community members already knew most people on our team. However, building rapport with the community during these meetings was essential for the group members who had not previously worked with Community A. Thus, time was spent introducing ourselves and our goals and conversing with the community members. To understand the community's goals for this project, it was essential to understand the context of the community, their values, beliefs, journeys, destinations, language, knowledge, and more through participant observations. We spent a lot of time trying to understand the knowledge community members had about C&DW

and recycling. We learned the community was already utilizing C&DW in their homes and on shared roads but in unsustainable ways. While there is a Junta de Acción Comunal for the community (a legally protected organized civil structure made up of members of the community), there are also natural leadership structures and leaders as well. Despite this divide in power structures, both groups wanted to find a way to make recycling C&DW profitable and beneficial for the community. The understanding of these relationships contributed to the understanding of urban mining potential in the community.

In Buenos Aires, jointly with the representatives of two local government agencies, three visits to e-waste facility A and two to facility B were conducted in August 2022. In facility A, the researcher was accompanied by a team consisting of one toxicologist and three social workers, and in facility B an environmental professional led the visit. Three other e-waste facilities in the province of Buenos Aires were visited by the researcher. During these encounters, neither video nor audio recordings were made. Photographs of the workspace, machinery, devices, and waste were taken with the previous authorization of participants. The objective of this observation was to understand the different contexts of e-waste workers, their work dynamics, power distributions, needs, concerns, and desires. These interactions helped identify new actors and refine data on materials and equipment, collection and treatment practices, and the value chain characteristics.

4.4. Workshops

Through the preliminary analysis of the data gathered on C&DW and Community A during interviews and participant observations, we formulated a common theme, which was the desire to bring educational opportunities to the community and conduct workshops to give community members, especially women, skills to generate income. To follow our commitment to a community-centered research approach, we developed a participatory workshop with Community A that took place in March 2023. This workshop presented an opportunity for women and low-income community members in Colombia, specifically targeted towards Community A in this approach, to engage one another in the process of learning about recycling concrete from C&DW and develop a plan for how this can become actionable in their community. The workshop had five key sections: C&DW Composition and Values, Environmental Aspects, and Necessary Permits; C&D Recycling Processes and Technologies; Occupational Health, and Safety; Applications and Entrepreneurism; and Pathways Forward. We worked with community members through virtual communications and surveys to ensure that the included components were necessary and relevant. Additionally, to centralize local knowledge and build local capacity we invited Colombian subject matter experts to lead each of the key sections.

For e-waste communities, two workshops were held with each group of workers from facilities A and B. The main objective of the workshops was to analyze workers' perceptions regarding the chemical risks related to their activity, train them in basic concepts of risk prevention and management, and collect their opinions on a proposed intervention to prevent the open burning of cables. All the e-waste workers at facility A (five males) participated in the workshop with their coordinator (one male). At site B, since the cooperative is not formally working with e-waste, the associated urban recyclers with e-waste experience were invited to participate. In total, 37 (17 females, and 20 males) and their coordinator (1 male) participated in the workshop. The activities were conducted in two hours and included: (1) Initial general risk identification activity, (2) E-waste risk perceptions activity, (3) Discussion about a cable stripper prototype, and (4) Community mapping of burning sites, metal buyers, and collection points (only at site A). Audio recordings were taken with the prior permission of the participants.

4.5. Research Extension with the Undergraduate Students of the Colorado School of Mines

Following the teaching philosophy of the Colorado School of Mines (CSM) Humanitarian Engineering and Science program [41], three graduate–undergraduate research extension activities were carried out with CSM undergraduate students. First, the topic

"Empowering People: Extracting Value from Waste Through Urban Mining" was proposed as project motivation in the "Design I" course in Fall 2022, aimed at more than 600 first-year engineering students. Second, for a senior design course, specific sociotechnical challenges related to C&DW and e-waste were proposed to the students. Third, a new version of the "Engineering and Sustainable Community Development" course was delivered to 24 undergraduate students in Spring 2023, based on three e-waste technical challenges defined by a community of recyclers in Bogotá, Colombia.

4.6. Preliminary Review of the Mining Engineering Curricula

To describe the approach of urban mining and inclusive urban mining as topics in the engineering curricula, we conducted preliminary research on curricular databases from universities in the United States. We selected the top universities in mining engineering, as listed on the National Mining Association website [42], and then we examined each university's website and online course catalog individually, solely looking at their minimum requirements for obtaining a Mining Engineering Bachelor of Science in the 2022–2023 academic year. University, general education, and B.S. course requirements were not included. When examining each catalog, we searched each course description to determine if urban mining concepts and sociotechnical approaches were explicitly stated as learning goals in required courses. To search for urban mining concepts, we utilized the following search terms: "urban mining", "construction and demolition waste", "C&DW", "electronic waste", and "e-waste". We also searched for sociotechnical learning approaches by searching for the terms "community", "sociotechnical", "social", "societal", or "human". Acknowledging the limitations of this approach because of its subjectivity related to the lack of detailed information describing the material reviewed within the required courses as well as the research projects being conducted within these universities external to course curricula, we incorporated our findings as a preliminary set of data that could be further analyzed in future works.

5. Inclusive Urban Mining in Latin America

Currently, in most Latin American countries, the "resource recoverers" [43] or, as we refer to them in this paper, "urban miners", are not yet integrated into the regulatory framework. Their working conditions are often precarious, exposing them to hazardous chemicals, including heavy metals and halogenated compounds [44]. Although they provide a critical environmental service, waste workers have historically been stigmatized and excluded within society [43].

With growing interest, but still nominal in comparison with traditional mining, some countries in the region are facing the challenge of regulating the activity of urban mining, integrating informal workers, and promoting improvements in processes and technologies to increase productivity and promote sustainable local economic development [35,36].

To help overcome the challenges and barriers enumerated, we present below the current and potential benefits of urban mining for communities and the environment with a special focus on Latin America. We support our claims by providing evidence from our literature research and experiences with communities in Argentina and Colombia interested in treating waste not only as a source of income but also as a way of empowerment. We introduce inclusive urban mining as a concept that has its roots in inclusive solid waste management, an activity with a long history in these two countries [45,46].

5.1. Why Should Urban Mining Be Inclusive? Some Examples of Its Social Benefits

5.1.1. Women's Empowerment: Through Recycling C&DW

When asked about their role in their community, household, and workplace, many of the women in Community A in the Colombia research claimed to be a leader of some kind. They either defined that as having a position of power and knowing they could tell others what to do, or being a mentor or a friend to people when they needed something. Some also cited their age when asked about their role, stating that because they are older, they are wiser and therefore better leaders. When asked about the problems women face in their community, problems with children were often cited, such as children being left alone or turning to illegal activities to make money. Another problem that often arose was unemployment, sometimes due to the lack of transportation or job opportunities. Finally, when asked about solutions to these problems and how the term women's empowerment was understood, people often brought up workshops that had been conducted in the community in the past. These workshops often focused on cosmetology, such as doing hair and nails, art practices, baking, or cooking. Many people mentioned education and the importance of learning, and some brought up making money and having the ability to secure and spend money for themselves or their family while in the confines of their own homes. When asked about women's empowerment, one leader in the community stated, "[Women's empowerment] is the idea that women can work on their own with their own capacities". She also discussed the importance of bringing opportunities to the community and doing workshops to give women tools to find jobs.

As illustrated in the semi-structured interviews conducted with women in Community A, in this context, it was found that urban mining can best empower them helping them generate more income and advance their education to gain additional skill sets. While financial and economic decision-making power is a common dimension of women's empowerment, the details of how exactly this pursuit could be more beneficial and empowering to women in Community A were developed through a dialogue and an understanding of the community context. For example, the women demonstrated the need for the time and capacity to care for their families alongside these pursuits, thus making this a homebound endeavor.

While urban mining shows promise to contribute to empowerment opportunities for multiple vulnerable groups, including women, it must be acknowledged that these contributions can be maximized through a contextual understanding of the complex systems that shape the lives of these groups. As such, there is a need for academic institutions, especially those related to engineering and design, to work alongside communities to understand how pursuits, such as urban mining, can empower them. Moreover, to ensure relevant empowerment to vulnerable groups, it is essential to take an interdisciplinary approach, as empowerment is contextually situated; thus, different ideas of empowerment exist within different contexts and are reflective of their own specific communities and cultures.

5.1.2. Social Transformation and Digital Inclusion in the E-Waste Sector

In our visits to e-waste facilities, we observed the pride of workers regarding their role as green actors in a context where circular economy policies and regulations, although necessary, are still pending. This role is one of their motivations when facing the many obstacles presented to them. To name a few, they have little bargaining power vis à vis buyers—mostly intermediaries—and limited access to information and technologies to maximize waste recovery. Even with all the challenges, these groups of workers, mostly born and raised in vulnerable conditions, go through a collective process of what they call "subsistence, resistance, and transformation". Some are young adults who have never kept a job for more than a couple of months, but in their cooperatives, they become resilient and learn not only specific knowledge relevant to their business, but also the general rules of the labor industry, such as complying with the schedule and attendance. Hence, as an interlocutor told us once, "We [the cooperative] not only recycle materials, but people". Therefore, it is not arbitrary that some cooperatives have included words, such as "dignity" or "justice" in their names. We see, then, that the feeling of belongingness that the activity generates in workers has the potential to contribute to the education on labor conduct, becoming a transforming process for specific groups, including young recyclers.

For workers in general, as in any other labor space, learning new skills and developing new knowledge are essential, and urban miners are not exempted from this process. They learn to repair and disassemble e-waste by gaining specific knowledge about electronics, IT, mechanics, and sometimes material composition and chemistry. However, particularly for this sector and especially in the Latin American context, learning these disciplines goes beyond training workers in their roles. This learning process also means a step towards their insertion in an increasingly demanding digital society. This is an additional benefit of urban mining, illustrated by the case of a worker who, during a workshop, told us, "Since they [the cooperative] gave me a computer, I was able to use one for the first time in 60 years". This worker's access to technology, although based on the objective of training him on electronics repairing, ended up meaning his access to a digital world that is often hampered for people his age.

In urban mining cooperatives, we have observed how not only materials but also people are transformed. Understanding this opens the way to many study areas that are scarcely explored today by the academic community. We question what other social benefits do urban mining cooperatives bring? Could the social benefits be externalities that account for the comparison between urban and traditional mining? We wonder, in particular, for the e-waste management sector, how could actors dedicated to digital inclusion and actors dedicated to e-waste management interact? What impacts would a more inclusive digital society have on the use, disposal, and management of electrical and electronic devices? Could the circular economy based on e-waste become a mechanism for digital inclusion?

5.1.3. The Value of Local Knowledge for the Global Development of Urban Mining: Examples from the E-Waste Sector

Although many of the e-waste recoverers did not perceive themselves as producers of knowledge or technology, as a result of our visits and workshops, we learned that the knowledge of these workers is as important as any other certified by an academic degree. For example, some workers can quickly identify components and materials with high efficiency, and some apply craft and ingenious low-cost plastic identification methods (e.g., by their texture, smell, or color). Others have perfected techniques, such as manual disassembly or burning, to improve the quality of the metals they obtain, even in conditions that create major health problems for them and their communities because of their exposure to hazardous chemicals [44]. Likewise, some workers with more than a decade of experience in the sector have undertaken the important task of sharing their knowledge with less experienced peers, providing in-person training and written material. The information is exchanged between the workers themselves. They themselves are the referents of the activity and share their knowledge. A good example is the free and public guidance document "Cooperación y reciclado para un mundo sustentable" ("Cooperation and recycling for a sustainable world") edited by Salcedo et al. in 2019 [47].

In the literature, waste workers are usually pigeonholed into informality [48], and under a global gaze that proposes external strategies to deal with local problems. We believe that to avoid the traditional labels of "lacking" or "informal," and put an end to the historical marginalization of the recyclers, waste pickers, and waste workers in general, it is necessary to study their resilient learning, improvement, and knowledge-transfer processes. Johansson et al. [18] claim that "the informal sector can nevertheless teach us how to change our perception of technospheric stocks and view them not as a problem but as a resource" (p. 42). We thus wonder how their voices could be amplified so that larger audiences know how their inventiveness and persistence can overcome the barriers of the context in which they work and how these skills and knowledges can help alleviate the global challenge of e-waste, C&DW, and other relevant waste streams. We then ask what can Latin American urban miners contribute to the global conversation on waste management? How can local knowledge improve foreign technological processes? These proposals are not in opposition, but on the contrary, they seek to promote a synergistic interaction between the development of knowledge and technologies in Latin American countries and countries of other regions.

5.2. Environmental Benefits of Urban Mining

Present-day demand for material resources combined with concerns about the sustainability of extraction practices and the effects of waste have increased the interest of both practitioners and scholars in the concept of the circular economy. In this context, urban mining is gaining momentum from various perspectives [49]. First, this practice rejects linear approaches to production, replacing the "end-of-life" stage of traditional waste management with reusing, recovering, and recycling processes throughout a product's life [50]. Second, many scholars agree that urban mining improves resource fulfillment by advancing the circular economy and minimizing environmental burdens [51]. Obtaining materials from discarded items can also contribute to climate change mitigation since metal recovery consumes less energy than the extraction of primary raw materials [51]. For instance, the energy needed for the manufacturing and transportation of building materials could be reduced by about 29% if these materials are recycled [7,52,53]. Third, urban mining provides a solution for uncontrolled waste management, which remains a significant global challenge [14] due to factors, such as the exposure of the environment and humans to hazardous substances and biological vectors. The accelerated growth of waste on a global scale results in valuable aboveground stocks in quantities that are often comparable to or exceed natural stocks [6]. For example, Grant et al. [3] indicate that "thirty smartphones contain as much gold as one ton of mine rock from a traditional gold mine" (p. 7). Thus, these resources have become attractive to those that acknowledge the gradual depletion of economically minable resources [49].

Environmental Benefits of Recycling C&DW in Colombia and E-Waste in Argentina

The construction industry is a main contributor to carbon dioxide emissions across the globe due to it containing many elements with high carbon footprints, such as cement and concrete production, transportation, and C&DW generation [54]. In 2020, the United Nations Environment Program (UNEP) stated that the buildings and construction sector accounted for 38% of the total global energy-related CO_2 emissions in 2019 [55]. The cement industry alone contributes to about 8% of the global CO_2 emissions [56]. Effectively managing C&DW is a critical component of preserving our environment, natural resources, economy, and society [57]. Despite this, C&DW mismanagement is a widespread issue.

Around the world, the problem of C&DW is worsening, thereby exacerbating environmental and social issues [58]. In Colombia, the expansion of the construction industry is aggravating these issues through the disposal of C&DW in an insufficient and unregulated manner and the increased illegal extraction of aggregate materials [59]. These increasing environmental and social issues are gaining national recognition in Colombia, as seen in Resolution 472, which outlines the management of C&DW in Colombia in light of the inadequate disposal and increased generation of C&DW in cities across the country, including Bogotá, Medellín, Santiago de Cali, Manizales, Cartagena, Pereira, Ibagué, Pasto, Barranquilla, Neiva, Valledupar and San Andrés [60] as well as other legislation released over the past couple years [61–63]. Recycling C&DW could contribute to reducing the inadequate and unregulated disposal of C&DW and decrease the illegal extraction of aggregate materials.

Regarding e-waste in Argentina, a national report estimates that 465,000 tons of this waste stream are generated per year [36] and only 4% is managed in an environmentally sound manner [35]. Roughly, following the methodology in Forti et al. [12], we calculated that this low percentage contributed to a net saving of 8 kt of CO_2, equivalent to emissions from the recycling of secondary raw materials substituted to virgin materials. If this percentage increases up to the goal of 30% under Target 3.2 of the ITU Connect 2030 Agenda [64], it might help save up to 60 kt of CO_2 equivalent emissions.

6. Inclusive Urban Mining in US Engineering Curricula

6.1. Preliminary Screening of Urban Mining Content in the US Engineering Programs

Table 1 summarizes the preliminary research findings on curricular databases from the universities in the United States with Mining Engineering Bachelor of Science programs.

Table 1. Preliminary review of Mining Engineering Curricula related to urban mining and sociotechnical learning approaches * in the US.

	Mining Engineering University												
Number of Required Courses	**1**	**2**	**3**	**4**	**5**	**6**	**7**	**8**	**9**	**10**	**11**	**12**	**13**
Urban Mining Concepts	0	0	0	0	0	0	0	0	0	0	0	0	0
Sociotechnical Concepts	0	2	0	1	0	1	0	4	3	1	0	0	0
Number of Non- Required Courses	**1**	**2**	**3**	**4**	**5**	**6**	**7**	**8**	**9**	**10**	**11**	**12**	**13**
Urban Mining Concepts	0	0	0	0	0	0	0	0	0	0	0	0	0
Sociotechnical Concepts	1	3	3	2	0	2	0	4	3	1	0	0	1

* Search terms: "urban mining", "construction and demolition waste", "C&DW", "electronic waste", "e-waste", "community", "sociotechnical", "social", "societal", "human".

There are 13 universities in the United States with Mining Engineering programs. Within these programs, urban mining was not explicitly listed as a learning goal in any course descriptions found within the universities' websites or online course catalogs depicting the minimum course requirements for obtaining a Mining Engineering Bachelor of Science in the 2022–2023 academic year. The programs include courses with sociotechnical approaches, or at least discuss human-based concepts to a certain extent; however, these approaches were found more often within the non-required courses.

To reiterate, our findings should be viewed as a preliminary set of data that could be further analyzed in future works due to the limitations of this approach, including its subjectivity related to the lack of detailed information describing the material reviewed within the required courses in course descriptions found on university websites. Additionally, this screening did not include research projects being conducted within these universities external to course curricula.

It is not the intention of this work to only focus on the US curricula but to call on global engineering and technology academic institutions to involve current waste management challenges in their programs as motivators for technological innovation projects. Through the examples described below, we want to emphasize that urban mining could be a significant research subject and an excellent educational opportunity for organizations specializing in traditional mining, as these organizations could take advantage of their existing technical knowledge in material extraction and processing.

6.2. *Approaches to Include Urban Mining in the US Engineering Curricula*

6.2.1. Introducing Urban Mining to First-Year Engineering Design Courses

The faculty in the design 1 and 2 courses at CSM are diverse in terms of academic and professional backgrounds. They mostly have STEM majors, such as design, civil engineering, electrical engineering, and other traditional engineering disciplines. Some have experience in the industry, and some others are senior researchers. This diversity gives the students exposure to different industries and "real-world problem-solving and design experiences", as our interviewee claimed. In total, they teach 25 groups of 25 first-year engineering students by applying an ambitious but enriching approach involving problem formulation, design thinking, and stakeholder engagement.

The purpose of their teaching approach is to expose students to stakeholders and communities they did not know about and to make them reflect on how engineering projects might affect those communities. Its purpose is also to broaden students' perspectives in ways they might never find out on the news or social media. This approach is not exempt from resistance, either from the faculty or students. According to the instructor, many students tend to separate the technical and the social and only focus on the technical challenge because, in the end, "the majority is going to end up in traditional engineering jobs one day, and that is just the path they want". Some other students understand the complex issues that low-income communities are facing, but they just do not want to get involved. In this context, the instructors try to emphasize the importance of integrating knowledge. They explain to students that engaging with stakeholders and considering the specific contexts, geographies, cultures, and expectations are key stages in the design process. "You cannot do technical in a vacuum", our interviewee claimed. As an example, he asks students, "Could you design a technical solution without considering government regulations"?

The efforts of going beyond the boundaries of traditional engineering are huge for the faculty, and despite some room for improvement in terms of genuine stakeholder engagement, those instructors with traditional engineering backgrounds are proud of what they are doing in this class. They are aware that even if it is an introductory course, they provide additional techniques that are not usually offered in the first year in other programs in the US.

At CSM, students receive a "call for proposal" (CFP) broad enough so they can elaborate on the problem after a series of research stages that can involve literature research and consultation with subject experts, potential users, and other stakeholders. In 2022, for the very first time, the CFP was developed in collaboration with two graduate students from the Humanitarian Engineering and Science program, who are the authors of this paper. The topic, "Empowering People: Extracting Value from Waste Through Urban Mining" was innovative since it introduced urban mining, life cycle, and waste management as motivations for design. The students received a brief description of the general situation of waste management systems in low-income communities and the specific challenges and opportunities in the study sites that we present in this paper.

The way the CFP was developed allowed students to set the boundaries of the problems by themselves, encouraging them to think creatively and "out of the box", to look at things outside their own immediate context, and to familiarize with the processes that happened after "the magic truck comes by and picks up the purple bin".

From a total of approximately 625 students, 110 students grouped in 25 teams achieved the 20% best-scored projects. Among the winning teams, the distribution of themes was Food/organic waste (5), E-waste (4), Plastic/packaging waste (4), Household effluents (2), Medical waste (2), Textile waste (2), and Others/Out of scope (6).

The CFP motivated students to reflect further down the line at the end of product life. For example, some students worked on recycling technologies to be applied locally. Others looked at extending the life cycle by redesigning products or tried to look for upcycling opportunities at the source. A number of students preferred to address the specific challenges related to the settings and contexts that we have presented. They usually choose their path according to what they are exposed to and tend to lean towards stakeholders that they already know. The groups interacted with recycling, electronics, processing companies, big warehouses, consumers, and professional users.

Although the experience was enriching for both students and professors, introducing a new way of mining provoked tensions in a school well-known for its mining tradition. Some mining professionals asked, "Why are they calling this mining"? and claimed, "This is not our definition of mining. This is not what mining engineering is". We wonder, then, what does it take for the traditional disciplines to extend their boundaries? In the end, changes in the curricula that in the past seemed far off, such as the inclusion of the social aspects in a first-year engineering design course, became a reality seven years ago. We wonder, what is urban mining lacking to be considered as a topic of relevance by the traditional mining sector? What are the differences? What are the convergence points?

6.2.2. Introducing Urban Mining in Elective Courses: The Case in an Engineering and Sustainable Community Development course

For the very first time, in Spring 2023, the course Engineering and Sustainable Community Development (ESCD) was taught in a project-based format, involving direct interaction with communities. This course gathered twenty-five third-year and graduate students to work collaboratively with a Colombian recycling association to improve three processes: e-waste plastic identification, copper separation from cables, and precious metals separation from circuit plastic boards. From the instructor's perspective, urban mining is not the initial motivation of students that join this course. These students care about community-based projects in general, and even if they have education in specific disciplines (Environmental Engineering, Civil Engineering, Mechanical Engineering, Chemical Engineering, and Design Engineering), they are curious about the different ESCD opportunities for practice. Addressing a waste-related topic and how other communities interact with it stimulates students to think in ways they never have.

According to the instructor, urban mining could be framed as the future of mining. He thinks it has the potential to convene students, researchers, and industry professionals who are not usually involved in traditional mining (for example, electrical and civil engineers) that would be able to apply their knowledge and skills to transform anthropogenic waste stocks into valuable materials. In this sense, traditional mining institutions could see urban mining as a way to expand their curricula, staff, and areas of expertise. The expansion, however, will not be easy for those that have a traditional mining background, he said. It will require not only their willingness but the comprehension of new knowledge to deal with mines not located in the mountains but in the cities. Therefore, the challenge ahead will be to deal with the technical differences as well as the intricate relationship between material extraction and urban systems.

When we asked him how to make urban mining visible and relevant, the instructor did not hesitate to claim that extending graduate research into first and third-year design courses is an important grassroots step that could eventually position the topic as an institutional priority from the top-down. He explained that improving urban mining not only contributes to the circular economy but also provides employment opportunities for marginalized groups of people, such as those displaced by violence, poverty, or climate change. In this sense, applying a community-based approach in the engineering curricula gains more significance when urban mining is seen as an employment solution. He acknowledged that engineers could address urban mining from an industrial, automated, and large-scale perspective, but in doing so, they might be ignoring and neglecting the current labor problems that cities are facing and the minor waste streams that are managed in small neighborhoods. "All those stocks of waste are always going to exist, and all those people needing employment are always going to exist irrespective of the big machinery that you put in place". Hence, he emphasizes the need for more engineers and engineering students to be trained to co-work with marginalized communities with the aim of improving their processes, products, and labor conditions.

The instructor also pointed out some important parallels between artisanal small-scale gold mining (ASGM) and urban mining. Less than a decade ago, the Minamata Convention forced countries to focus on reducing and, when feasible, eliminating the use of mercury. ASGM became a major area of interest for many institutions, including well-known traditional mining schools and research centers in the Global North. This new area of interest opened opportunities for research and practice in fields, such as engineering and social sciences. The recurrent presence of informality and the way in which communities engage in these activities, sometimes ending up exposing them and their families to hazardous chemicals, are other points of commonality between ASGM and urban mining. Furthermore, ASGM and urban miners both "are for the most part invisible to mainstream society", the professor said. Taking this into account, we wonder if similar drivers, such as the global concern about scarce materials, including rare earth elements, metals, and minerals, might have the same result for urban mining. Will inclusive urban mining become a field of research and practice in the way that artisanal scale gold mining did, even with the tensions and resistance that it generates?

6.2.3. Introducing Urban Mining Projects in Third and Fourth-Year Project-Based Design Courses

In an effort to offer the upper-level engineering students opportunities to learn more about human-centered design and humanitarian engineering challenges, CSM offers a three-semester hour project-based design course targeted towards junior- and senior-level students. Within the Fall 2022 semester, multiple graduate students from the Humanitarian Engineering and Science program—including the authors of this paper—were able to work with project teams in this course on specific real problems affecting real people. Overall, of the 23 students registered for this course in the Fall 2022 semester, 10 students worked on urban mining-related projects. One group of four students worked on a C&DW-related project, while two groups of three students were devoted to e-waste-related projects.

The faculty member responsible for facilitating the course in the Fall 2022 semester described urban mining as an opportunity, not only to "emphasize reclamation of precious materials in environmentally friendly ways that are also economically beneficial to disadvantaged populations" but to push back on the negative connotation associated with the term "mine" due to the often-harmful activities, practices, and ramifications of the industry. The professor believes urban mining is a way to "reclaim the word 'mine' for positive applications", and institutions responsible for the progression of the often damaging activities, practices, and ramifications of traditional mining processes, such as universities including CSM, should be at the forefront of developing "more environmentally friendly and socially equitable ways of mining and engineering", such as urban mining. The inclusion of urban mining in the curricula to atone for the negative externalities involved in traditional mining can also be an opportunity to enhance mining engineering education by facilitating understandings of concepts, such as life cycle analysis.

In addition, the professor stressed the importance of understanding the local context of projects, such as the cultural, socioeconomic, and environmental dimensions of the cities with which they work, and utilizing approaches from the social sciences and environmental sciences to develop solutions that are "most appropriate to their target population and do the least harm to the same population as well as their environment". He stated that this utilization and understanding was even more essential than technical foundations, such as the engineering mindset, to arrive at a point where the technical solutions were appropriate. Furthermore, to develop the best solutions possible, the professor argued that stakeholder engagement, particularly empathetic stakeholder engagement ("which is culturally sensitive, appropriate for local contexts, aware of potential unintended consequences, and ultimately in search of the greatest number of 'win-win' situations as possible, where the environment is also a key stakeholder"), is essential. We the question how critical social and environmental science approaches can have a space in engineering curricula, as these topics (particularly the social sciences), despite their importance, are traditionally shunned in engineering education.

7. Limitations of This Study

There are two major limitations in this study that could be addressed in future research. The first limitation is related to the number and selection of participants included in our interviews and workshops, which were relevant in terms of the qualitative analysis of the C&DW and e-waste contexts in Colombia and Argentina presented but not statistically representative for a quantitative analysis. The second limitation is related to the reduced scope of our curricular review, since courses in departments other than Mining Engineering should be explored, including Chemical Engineering, Environmental Engineering, Resources Engineering, and Materials Engineering, among others.

8. Conclusions

As illustrated above, urban mining has a leading role in the circular economy that is currently developing. However, in particular contexts, such as in Latin American countries, this activity poses additional benefits that can be maximized if they are understood and studied. We state that the study of urban mining from an interdisciplinary approach could contribute to this field in order to achieve a much more inclusive and sustainable activity.

For urban mining, cities are the locations where the extraction, circulation, and accumulation of materials take place. Hence, to favor inclusive urban mining it is not only necessary to understand collection and extraction processes but also to understand cities and their context. Contextualizing this activity means analyzing local legal frameworks, stakeholders involved, their history, ideologies, culture, alliances and power differentials, the flow of materials, current technologies, and processes. In light of our findings, we argue that the future challenges associated with inclusive urban mining are sociotechnical in nature. Thus, we highlight the importance of promoting community-based research methods and concepts from the Engineering and Sustainable Community Development practices [65] to be included in mining, materials, metallurgical science, and engineering academic programs as a way to address these challenges.

It is not without reason that we have argued the case for working alongside communities to solve problems in a participatory way, as the knowledge of the local community members is crucial for developing sustainable and just solutions. However, this effort makes it necessary to promote knowledge sharing throughout the entire problem-solving process within and between multiple fields, disciplines, and communities and it also exemplifies the importance of fostering sustainable networking pathways. Productive interactions between groups are fundamental to maximizing the capacity to collaboratively find a viable, just, and long-term solution to community problems. To understand effective knowledge sharing, we argue that studying groups that are already doing this successfully is essential. Uniminuto, a Colombian University, is a prime example of an academic institution striving to create a positive social impact and uplift the vulnerable communities. Through knowledge and experience gained in PCIS projects, they have developed the "Social Innovation Route Framework" [40], a five-stage framework outlining community engagement projects, which is a powerful tool that academic institutions, especially those related to engineering and design, can utilize to take a proactive role in (1) working with vulnerable groups to improve their labor and environmental conditions and (2) understanding the sociotechnical dimensions of their projects.

Our observations also reflect the benefits of educational proposals that combine engineering knowledge with concepts from the sustainable community development framework, which is based on the social sciences. Thus, the interdisciplinary approaches that motivate students to make a contextual analysis of their projects, including history, politics, ideology, ethics, and culture, influence the way in which they develop their inventions. These approaches could also bring them closer to the Latin American context without falling into methodologies of the North–South dominance.

To answer the questions that were raised in this work, we propose some additional areas for future research. First, there is a need for a deeper analysis of the US and Latin American science engineering curricula to understand the lack of urban mining content and identify synergistic opportunities with overseas academic entities. Second, further work needs to be undertaken to screen current educational programs in the US since our study is limited. Third, future efforts within the engineering education field should be focused on developing an outline of an Inclusive Urban Mining course with the insights from this research that could be then included in the US engineering programs. Fourth, additional topics that should supplement the study of inclusive urban mining should be identified. Some topics that we believe could be beneficial to include are understanding the material politics of what is traditionally viewed as "waste" as well as learning social science approaches, such as contextual and empathetic stakeholder engagement strategies, to properly understand cities and their context.

Author Contributions: Conceptualization, S.L.S. and J.E.S.; Data curation, S.L.S. and J.E.S.; Formal analysis, S.L.S. and J.E.S.; Funding acquisition, S.L.S. and J.E.S.; Investigation, S.L.S. and J.E.S.; Methodology, S.L.S. and J.E.S.; Project administration, S.L.S. and J.E.S.; Resources, S.L.S. and J.E.S.; Validation, S.L.S. and J.E.S.; Writing—original draft, S.L.S. and J.E.S.; Writing—review and editing, S.L.S. and J.E.S. All authors have read and agreed to the published version of the manuscript.

Funding: The studies were funded by the Colorado School of Mines (Mines) Humanitarian Engineering and Science Program (Humanitarian Engineering & Science Grant 2022).

Institutional Review Board Statement: Both projects were approved by the Colorado School of Mines Human Subjects Research Team and exempted from the Institutional Review Board (IRB) process requirements.

Informed Consent Statement: Informed consent was obtained from all participants involved in both studies.

Data Availability Statement: Some data presented in this study are available on request from the corresponding authors. These data are not publicly available to protect the identity of participants. Other data presented in this study are available in the sources listed in References.

Acknowledgments: The authors would like to acknowledge and express their gratitude to the Colorado School of Mines (Mines) Humanitarian Engineering and Science Program for their support, the Corporación Universitaria Minuto de Dios (Uniminuto) teams that supported the Colombia project, the Fulbright Commission and the Ministry of Education in Argentina that supported one of the researchers' education, and the members of e-waste programs and cooperatives that contributed with their time, patience, and knowledge. Any opinions, findings, and conclusions or recommendations expressed in this material are those of the author(s) and do not necessarily reflect the views of the supporting organizations. The authors would also like to acknowledge Juan Lucena (Mines) for his tireless work, advising the authors and continuously presenting us with opportunities for our personal and professional development and Martha Liliana Herrera Gutiérrez (Uniminuto), who was in charge of translation, facilitation, and communication in the Colombia project. We express our gratitude to the community members in Argentina and Colombia and the interviewees that made this research possible.

Conflicts of Interest: The authors declare that they have no competing financial interest or personal relationship that could have appeared to influence any part of their work as reported in the paper. The funders had no role in the design of the study; in the collection, analyses, or interpretation of data; in the writing of the manuscript; or in the decision to publish the results.

References

1. Jacobs, J. *The Economy of Cities*; Random House: New York, NY, USA, 1969; ISBN 0-394-70584-X.
2. Aldebei, F.; Dombi, M. Mining the Built Environment: Telling the Story of Urban Mining. *Buildings* **2021**, *11*, 388. [CrossRef]
3. Grant, R.J.; Oteng-Ababio, M. The Global Transformation of Materials and the Emergence of Informal Urban Mining in Accra, Ghana. *Afr. Today* **2016**, *62*, 3–20. [CrossRef]
4. Nakamura, T.; Halada, K. Potential of Urban Mine. In *Urban Mining Systems*; Springer Briefs in Applied Sciences and Technology; Nakamura, T., Halada, K., Eds.; Springer: Tokyo, Japan, 2015; pp. 7–29. ISBN 978-4-431-55075-4.
5. Chand, S.; Ranjan Rout, P.; Pathak, P. Basic Concepts, Potentials, and Challenges of Urban Mining. In *Urban Mining for Waste Management and Resource Recovery: Sustainable Approaches*; Pathak, P., Ranjan Rout, P., Eds.; CRC Press, Taylor & Francis Group: Boca Raton, FL, USA, 2022; pp. 1–11. ISBN 978-1-00-320107-6.
6. Cossu, R.; Williams, I.D. Urban Mining: Concepts, Terminology, Challenges. *Waste Manag.* **2015**, *45*, 1–3. [CrossRef] [PubMed]
7. Ghisellini, P.; Ncube, A.; Casazza, M.; Passaro, R. Toward Circular and Socially Just Urban Mining in Global Societies and Cities: Present State and Future Perspectives. *Front. Sustain. Cities* **2022**, *4*, 1–14. [CrossRef]
8. Maes, T.; Preston-Whyte, F. E-Waste It Wisely: Lessons from Africa. *SN Appl. Sci.* **2022**, *4*, 72. [CrossRef]
9. Ntapanta, S.M. 'Lifescaping' Toxicants: Locating and Living with e-Waste in Tanzania. *Anthropol. Today* **2021**, *37*, 7–10. [CrossRef]
10. Herat, S. Sustainable Management of Electronic Waste (e-Waste). *Clean* **2007**, *35*, 305–310. [CrossRef]
11. Lundgren, K.; International Labour Office; Programme on Safety and Health at Work and the Environment; Sectoral Activities Department. *The Global Impact of E-Waste: Addressing the Challenge*; International Labour Organization: Geneva, Switzerland, 2012; ISBN 978-92-2-126897-0.
12. Forti, V.; Baldé, C.P.; Kuehr, R.; Bel, G. *The Global E-Waste Monitor 2020: Quantities, Flows and the Circular Economy Potential*; United Nations University (UNU): Bonn, Germany; United Nations Institute for Training and Research (UNITAR)-co-hosted SCYCLE Programme: Geneva, Switzerland; International Telecommunication Union (ITU) & International Solid Waste Association (ISWA): Rotterdam, The Netherlands, 2021; ISBN 978-92-808-9114-0.
13. Colorado, H.A.; Muñoz, A.; Neves Monteiro, S. Circular Economy of Construction and Demolition Waste: A Case Study of Colombia. *Sustainability* **2022**, *14*, 7225. [CrossRef]
14. Margallo, M.; Ziegler-Rodriguez, K.; Vázquez-Rowe, I.; Aldaco, R.; Irabien, Á.; Kahhat, R. Enhancing Waste Management Strategies in Latin America under a Holistic Environmental Assessment Perspective: A Review for Policy Support. *Sci. Total Environ.* **2019**, *689*, 1255–1275. [CrossRef]

15. Moreno, P. State of Play for Circular Built Environment in Latin America and the Caribbean. A report compiling the regional state of play for circularity in the built environment in LAC across Argentina, Brazil, Chile, Cuba, Ecuador and Mexico. Ecosur and United Nations One Planet Network Sustainable Buildings and Construction Programme. 2020, pp. 1–34. Available online: https://www.oneplanetnetwork.org/knowledge-centre/resources/state-play-circular-built-environment-latin-america-and-caribbean (accessed on 28 February 2023).
16. Xavier, L.H.; Ottoni, M.; Abreu, L.P.P. A Comprehensive Review of Urban Mining and the Value Recovery from E-Waste Materials. *Resour. Conserv. Recycl.* **2023**, *190*, 106840. [CrossRef]
17. Arain, A.L.; Neitzel, R.L.; Nambunmee, K.; Hischier, R.; Jindaphong, S.; Austin-Breneman, J.; Jolliet, O. Material Flow, Economic and Environmental Life Cycle Performances of Informal Electronic Waste Recycling in a Thai Community. *Resour. Conserv. Recycl.* **2022**, *180*, 106129. [CrossRef]
18. Johansson, N.; Krook, J.; Eklund, M.; Berglund, B. An Integrated Review of Concepts and Initiatives for Mining the Technosphere: Towards a New Taxonomy. *J. Clean. Prod.* **2013**, *55*, 35–44. [CrossRef]
19. Universidad El Bosque. *Regional Electronics Study and Circularity in the America and the Caribbean (LAC) Region. Mapping of Stakeholders*; United Nations Environment Programme (UNEP), Global Environment Facility (GEF), SAICM Secretariat: Bogotá, Colombia, 2021; pp. 1–37.
20. Zeng, X.; Xiao, T.; Xu, G.; Albalghiti, E.; Shan, G.; Li, J. Comparing the Costs and Benefits of Virgin and Urban Mining. *J. Manag. Sci. Eng.* **2022**, *7*, 98–106. [CrossRef]
21. Adach-Pawelus, K.; Gogolewska, A.; Górniak-Zimroz, J.; Kiełczawa, B.; Krupa-Kurzynowska, J.; Paszkowska, G.; Szyszka, D.; Worsa-Kozak, M.; Woźniak, J. A New Face of Mining Engineer—International Curricula to Sustainable Development and Green Deal (A Case Study of the Wrocław University of Science and Technology). *Sustainability* **2021**, *13*, 1393. [CrossRef]
22. Smith, J.; Lucena, J.; Rivera, A.; Phelan, T.; Smits, K.; Bullock, R. Developing Global Sociotechnical Competency Through Humanitarian Engineering: A Comparison of In-Person and Virtual International Project Experiences. *J. Int. Eng. Educ.* **2021**, *3*, 1–29.
23. Lucena, J. Can Community Development Projects in Engineering Education Be Both Responsible and Sustainable? Theory, Education, and Praxis. In Proceedings of the 2020 ASEE Virtual Annual Conference Content Access Proceedings, Virtual, 19 June 2020; p. 34254.
24. Internal Displacement Monitoring Centre (IDMC). *Handbook for Developing National ASGM Formalization Strategies within National Action Plans*; Norwegian Refugee Council: Geneva, Switzerland, 2018.
25. DeJong, T. Desk Review of Artisanal and Small-Scale Gold Mining (ASGM) in Burkina Faso. Wash. DC USAID Artis. Min. Prop. Rights Task Order Strength. Tenure Resour. Rights II STARR II IDIQ 2019. Available online: https://www.land-links.org/wp-content/uploads/2019/10/USAID-AMPR-Burkina-Faso-ASGM-Desk-Review_FINAL.pdf (accessed on 28 February 2023).
26. Cazabat, C.; O'Connor, A. *2021 Internal Displacement Index Report*; Internal Displacement Monitoring Centre: Geneva, Switzerland, 2021; pp. 1–83.
27. Carrillo, A.C. Internal Displacement in Colombia: Humanitarian, Economic and Social Consequences in Urban Settings and Current Challenges. *Int. Rev. Red Cross* **2009**, *91*, 527–546. [CrossRef]
28. Ojeda, D. Landslides in Colombia and Their Impact on Towns and Cities. In Proceedings of the 10th congress of the International Association for Engineering Geology and the Environment (IAEG), Nottingham, UK, 6–10 September 2006; The Geological Society of London: Nottingham, UK, 2006; Volume 112, pp. 1–13.
29. Yuan, H.; Shen, L. Trend of the Research on Construction and Demolition Waste Management. *Waste Manag.* **2011**, *31*, 670–679. [CrossRef]
30. Pacheco Bustos, C.A.; Fuentes Pumarejo, L.G.; Sánchez Cotte, É.H.; Rondón Quintana, H.A.; Pacheco Bustos, C.A.; Fuentes Pumarejo, L.G.; Sánchez Cotte, É.H.; Rondón Quintana, H.A. Residuos de construcción y demolición (RCD), una perspectiva de aprovechamiento para la ciudad de barranquilla desde su modelo de gestión. *Ing. Desarro.* **2017**, *35*, 533–555. [CrossRef]
31. Suárez-Silgado, S.S.; Betancourt-Quiroga, C.; Molina-Benavides, J.; Mahecha-Vanegas, L. La gestión de los residuos de construcción y demolición en Villavicencio: Estado actual, barreras e instrumentos de gestión. *Entramado* **2019**, *15*, 224–244. [CrossRef]
32. Menegaki, M.; Damigos, D. A Review on Current Situation and Challenges of Construction and Demolition Waste Management. *Curr. Opin. Green Sustain. Chem.* **2018**, *13*, 8–15. [CrossRef]
33. García Botero, L.M. Sostenibilidad de la Disposición de Escombros de Construcción y Demolición en Bogotá. Master's Thesis, Civil Engineering, Universidad de Los Andes, Bogotá, Colombia, 2003.
34. Mendez-Fajardo, S. Aprovechamiento de Escombros: Una Oportunidad Para Mejorar La Infraestructura de Las Comunidades Marginadas. In Proceedings of the II Conferencia Internacional "Gestión de Residuos en América Latina, GRAL", Cali, Colombia, 15–17 June 2011; Universidad Autónoma de Occidente, Universidad Autónoma del Valle, International Waste Working Group (IWWG), UNICEF: Cali, Colombia, 2011; pp. 1–9.
35. Wagner, M.; Baldé, C.P.; Luda, V.; Nnorom, I.C.; Kuehr, C.; Iattoni, G. *Regional E-Waste Monitor for Latin America: Results for the 13 Countries Participating in Project UNIDO-GEF 5554*; UNU/UNITAR: Bonn, Germany, 2022.
36. Maffei, L.; Di Santo, C.; Martínez, L.; Rodríguez, E.; Burucua, A.; Iribarne, R.; Malalán, T. Gestión Integral de RAEE. In *Los Residuos de Aparatos Eléctricos y Electrónicos, una Fuente de Trabajo Decente Para Avanzar Hacia la Economía Circular*; Ministerio de Ambiente y Desarrollo Sostenible de la Nación, Oficina de País de la Organización Internacional del Trabajo para Argentina: Buenos Aires, Argentina, 2020; ISBN 978-987-47600-3-6.

37. Maffei, L.; Burucua, A. *Residuos de Aparatos Eléctricos y Electrónicos (RAEE) y Empleo En La Argentina*; Oficina de País de la Organización Internacional del Trabajo para Argentina: Buenos Aires, Argentina, 2020; ISBN 978-92-2-031930-7.
38. International Labour Organization (ILO). Recommendation No. 193-Promotion of Cooperatives Recommendation. 2002. Available online: https://www.ilo.org/Dyn/Normlex/En/f?P=NORMLEXPUB:12100:0::NO::P12100_ILO_CODE:R193 (accessed on 1 March 2023).
39. Ottinger, G. Rupturing Engineering Education: Opportunities for Transforming Expert Identities through Community-Based Projects. In *Technoscience and Environmental Justice: Expert Cultures in a Grassroots Movement*; The MIT Press: Cambridge, MA, USA, 2011; pp. 1–312. ISBN 978-0-262-01579-0.
40. Pacheco Duarte, J.F.; Rodríguez Pupo, E.S.; Galindo Gómez, S.F. Ruta de innovación social: Paso a paso para desarrollar innovaciones sociales. Available online: https://hdl.handle.net/10656/11502 (accessed on 1 May 2022).
41. Lucena, J.; Chapman, E.; Rojas, M.; Schlezak, S.; Styer, J. Bringing It down from the Ivory Tower: Translating Engineering-for-Community-Development (ECD) Graduate Student Research into Community Engagement and Undergraduate Student Learning. In Proceedings of the ASEE 2022 Annual Conference, Minneapolis, MN, USA, 26–29 June 2022; American Society for Engineering Education: Washington, DC, USA, 2022; pp. 1–30.
42. US National Mining Association. Mining Schools and Universities. Available online: https://nma.org/about-nma-2/resources/mining-schools-and-universities/ (accessed on 1 January 2023).
43. Gutberlet, J. Solidarity Economy and Recycling Co-Ops in São Paulo: Micro-Credit to Alleviate Poverty. *Dev. Pract.* **2009**, *19*, 737–751. [CrossRef]
44. Deng, J.; Guo, J.; Zhou, X.; Zhou, P.; Fu, X.; Zhang, W.; Lin, K. Hazardous Substances in Indoor Dust Emitted from Waste TV Recycling Facility. *Environ. Sci. Pollut. Res.* **2014**, *21*, 7656–7667. [CrossRef] [PubMed]
45. Gutberlet, J.; Carenzo, S. Waste Pickers at the Heart of the Circular Economy: A Perspective of Inclusive Recycling from the Global South. *Worldw. Waste J. Interdiscip. Stud.* **2020**, *3*, 6. [CrossRef]
46. Gutberlet, J. Empowering Collective Recycling Initiatives: Video Documentation and Action Research with a Recycling Co-Op in Brazil. *Resour. Conserv. Recycl.* **2008**, *52*, 659–670. [CrossRef]
47. Salcedo, G.A.; Felicetti, R.; Pelloli, C.G. *Cooperación y Reciclado Para Un Mundo Sustentable*; CETCOT—Centro de Transferencia de Conocimiento y Tecnología; Universidad de Buenos Aires: Buenos Aires, Argentina, 2019.
48. Carenzo, S. Invisibilized Creativity: Sociogenesis of an "Innovation" Process Developed by Cartoneros for Post-Consumption Waste Recycling. *Int. J. Eng. Soc. Justice Peace* **2017**, *5*, 30–49. [CrossRef]
49. Koutamanis, A.; van Reijn, B.; van Bueren, E. Urban Mining and Buildings: A Review of Possibilities and Limitations. *Resour. Conserv. Recycl.* **2018**, *138*, 32–39. [CrossRef]
50. Kirchherr, J.; Reike, D.; Hekkert, M. Conceptualizing the Circular Economy: An Analysis of 114 Definitions. *Resour. Conserv. Recycl.* **2017**, *127*, 221–232. [CrossRef]
51. Lopes dos Santos, K. Unequal Geographies of Urban Mining: E-Waste Management in London, Sao Paulo and Accra. *Environ. Plan. Nat. Space* **2022**, *6*, 1–15. [CrossRef]
52. Thormark, C. The Effect of Material Choice on the Total Energy Need and Recycling Potential of a Building. *Build. Environ.* **2006**, *41*, 1019–1026. [CrossRef]
53. Blengini, G.A. Life Cycle of Buildings, Demolition and Recycling Potential: A Case Study in Turin, Italy. *Build. Environ.* **2009**, *44*, 319–330. [CrossRef]
54. Akhtar, A.; Sarmah, A.K. Construction and Demolition Waste Generation and Properties of Recycled Aggregate Concrete: A Global Perspective. *J. Clean. Prod.* **2018**, *186*, 262–281. [CrossRef]
55. UNEP. Building Sector Emissions Hit Record High, but Low-Carbon Pandemic Recovery Can Help Transform Sector-UN Report. Available online: http://www.unep.org/news-and-stories/press-release/building-sector-emissions-hit-record-high-low-carbon-pandemic (accessed on 28 February 2022).
56. Andrew, R. Global CO_2 Emissions from Cement Production. *Earth Syst. Sci. Data* **2018**, *10*, 195–217. [CrossRef]
57. Kabirifar, K.; Mojtahedi, M.; Wang, C.; Tam, V.W.Y. Construction and Demolition Waste Management Contributing Factors Coupled with Reduce, Reuse, and Recycle Strategies for Effective Waste Management: A Review. *J. Clean. Prod.* **2020**, *263*, 121265. [CrossRef]
58. Francisco, J.T.M.; De Souza, A.E.; Teixeira, S.R. Construction and Demolition Waste in Concrete: Property of Pre-Molded Parts for Paving. *Cerâmica* **2019**, *65*, 22–26. [CrossRef]
59. Bravo-German, A.M.; Bravo-Gómez, I.D.; Mesa, J.A.; Maury-Ramírez, A. Mechanical Properties of Concrete Using Recycled Aggregates Obtained from Old Paving Stones. *Sustainability* **2021**, *13*, 3044. [CrossRef]
60. Ministerio de Ambiente y Desarrollo Sostenible de Colombia Resolución No. 472. 2017. Available online: https://www.icbf.gov.co/cargues/avance/docs/resolucion_minambienteds_0472_2017.htm (accessed on 28 February 2023).
61. Ministerio de Ambiente y Desarrollo Sostenible de Colombia Resolución No. 1257. 2021. Available online: https://www.minambiente.gov.co/documento-normativa/resolucion-1257-de-2021/ (accessed on 28 February 2023).
62. Ministerio del Medio Ambiente de Colombia Resolución No. 541. 1994. Available online: https://minvivienda.gov.co/normativa/resolucion-0541-1994 (accessed on 28 February 2023).
63. Granados Santos, J.A.; Perez Alvarez, N.B.; Poveda Zambrano, A. Gestión integral de los residuos sólidos en el departamento de Cundinamarca. Available online: https://www.car.gov.co/uploads/files/5e29f9d0e2cda.pdf (accessed on 28 February 2022).

64. International Telecommunication Union (ITU). Connect 2030-An Agenda to Connect All to a Better World. Available online: https://www.itu.int:443/en/mediacentre/backgrounders/Pages/connect-2030-agenda.aspx (accessed on 1 January 2023).
65. Lucena, J.; Schneider, J.; Leydens, J.A. *Engineering and Sustainable Community Development*; Baillie, C., Ed.; Morgan & Claypool: San Rafael, CA, USA, 2010.

mining

MDPI

Article

Sociotechnical Undergraduate Education for the Future of Natural Resource Production

Jessica Smith [1,*], Carrie McClelland [1] and Oscar Jaime Restrepo [2]

1 Engineering, Design & Society, Colorado School of Mines, Golden, CO 80401, USA; cmclell@mines.edu
2 School of Mines, Universidad Nacional de Colombia, Medellín 050010, Colombia; ojrestre@unal.edu.co
* Correspondence: jmsmith@mines.edu; Tel.: +1-(303)-273-3944

Abstract: The greatest challenges for contemporary and future natural resource production are sociotechnical by nature, from public perceptions of mining to responsible mineral supply chains. The term sociotechnical signals that engineered systems have inherent social dimensions that require careful analysis. Sociotechnical thinking is a prerequisite for understanding and promoting social justice and sustainability through one's professional practices. This article investigates whether and how two different projects enhanced sociotechnical learning in mining and petroleum engineering students. Assessment surveys suggest that most students ended the projects with greater appreciation for sociotechnical perspectives on the interconnection of engineering and corporate social responsibility (CSR). This suggests that undergraduate engineering education can be a generative place to prepare future professionals to see how engineering can promote social and environmental wellbeing. Comparing the different groups of students points to the power of authentic learning experiences with industry engineers and interdisciplinary teaching by faculty.

Keywords: engineering education; sociotechnical thinking; global sociotechnical competency; corporate social responsibility; sustainability in mining engineering education

Citation: Smith, J.; McClelland, C.; Restrepo, O.J. Sociotechnical Undergraduate Education for the Future of Natural Resource Production. *Mining* **2023**, *3*, 387–398. https://doi.org/10.3390/mining3020023

Academic Editor: Mostafa Benzaazoua

Received: 12 March 2023
Revised: 25 May 2023
Accepted: 16 June 2023
Published: 18 June 2023

1. Introduction

The future of mining and other natural resource industries will require engineers who can take a sociotechnical approach to the challenges they face and the decisions they make in their working lives. The term sociotechnical recognizes that issues that appear to be technical in nature have an inherent social dimension [1–3]. For example, the coming energy transition will require massive amounts of minerals and metals, from the copper and iron necessary for power generation, transportation, and use, to the lithium, cobalt, and nickel required for electric vehicles. A recent review of around 30 energy transition minerals found that more than half are located close to vulnerable communities, specifically "on or near the lands of Indigenous and peasant peoples, two groups whose rights to consultation and free prior informed consent are embedded in United Nations declarations" [4]. Many of these populations already experience significant social and environmental injustices, and efforts to fast track mining projects central to the energy transition run the risk of adding more burdens [5]. Designing or selecting a particular technology that relies on one of these minerals, therefore, also implicates an entire supply chain of people, places, and injustices. To support more sustainable and responsible natural resource production, engineers need to be able to evaluate technologies and materials from a more holistic viewpoint, beyond narrow technical and economic considerations.

In general, there is a lack of preparation among engineering students to face the increasingly complex sociotechnical challenges of contemporary natural resource production. An ethnographic study of engineers practicing in the mining and oil and gas industries found that all had encountered sociotechnical challenges in their work, but felt underprepared to manage them [6,7]. The engineers described learning how to manage conflict, for example, as a trial by fire in which they learned on the fly, experimenting as they went along. This under-preparation largely stems from the structure of undergraduate engineering curricula in the United States–and likely elsewhere–placing a heavy emphasis on technical training, with very few opportunities for students to learn about the inherent social dimensions of the industries and infrastructures that will form the contours of their work [8]. Majors and courses frequently create an artificial technical/social dualism [9] that defines social and political concerns as external to engineers' domain [10].

Our educational research aims to address these structural challenges by nurturing sociotechnical thinking among engineering undergraduate students. We build on field evidence from the projects we have carried out in our countries, particularly in the United States and Colombia, in the teaching of mining and petroleum engineering, and put these into conversation with a broader movement of engineering educators seeking to disrupt the social/technical dualism by positioning engineering as "both technical and non-technical (taken to refer to the social, economic, political, ethical, etc.) from the start" [11]. Sociotechnical thinking involves students recognizing the "interplay between relevant social and technical factors in the problem to be solved" [8] and "to identify and address issues with an understanding of the complex ways in which the social and technical aspects of these issues are interconnected" by "holding both the technical and the social in one's mind simultaneously" [12]. Sociotechnical thinking is a prerequisite for students to be able to understand and promote social justice and sustainability through their professional practices.

Our research shows the value of using a sociotechnical perspective to teach and learn about themes related to sustainability and social responsibility. Engineering education research around the world assesses students' learning about these themes, though this field is too vast and growing too quickly to summarize here. The contributors to a special issue of the journal Sustainability focused on "Innovation in Engineering Education for Sustainable Development" (Sánchez-Carracedo 2020) captures current research in this area. A study of the International Center for Engineering Education in China found promise in its governance techniques to promote engineering education for sustainable development (Chen et al. 2022). A study of water and environmental engineers in Finland found correspondence between their sustainability education and required work skills, underlining the potential for engineers to play central roles in promoting sustainability (Vehmaa 2018). A group of US graduate students who traveled to India to study electronics manufacturing ended the trip reflecting on the ethical dimensions of regulations, gender roles, resources, and waste, including tensions among their multiple responsibilities (Berdanier 2018). Rich in-class discussions can also enhance student engagement and receptivity to sociotechnical thinking, especially given that the open-ended nature of these themes can prompt resistance (Blacklock et al. 2021).

In this article, we share the results of two efforts to nurture sociotechnical thinking among engineering undergraduate students. The first focuses on petroleum engineers at the Colorado School of Mines (Mines), where we integrated a critical approach to corporate social responsibility into multiple places in the curriculum. The second focuses on the Responsible Mining and Resilient Communities project that brought together engineering students from Mines, the United States Air Force Academy (USAFA), the University of Texas at Arlington (UTA), and the Universidad Nacional de Colombia-Medellín (UNAL). These students came from multiple disciplines but were all focused on artisanal and small-scale gold mining. We describe our methods and results case by case.

2. Materials and Methods

2.1. Sociotechnical Approaches to Corporate Social Responsibility in Petroleum Engineering

Our efforts to cultivate sociotechnical thinking in the undergraduate petroleum engineering program at Mines were part of a multi-year "Ethics of Extraction" research project, funded by the US National Science Foundation, that investigated the intersection of engineering and corporate social responsibility (CSR) [6]. Teaching students to recognize the inherent CSR dimensions of their work as engineers required taking a sociotechnical approach to both engineering, which is often viewed as a "technical" endeavor, and CSR, which can be viewed as a "social" endeavor. The type of CSR we taught was what Auld et al. [13] refer to as "new CSR," which encompasses activities that change core business practices to create social, economic, and environmental value for stakeholders as well as companies, in contrast with "old" CSR that is grounded in philanthropy. Changing core business practices in the mining and natural resource industries necessarily involves shifting engineering mindsets and practices.

For the project, we created an original survey instrument and custom enhancements for courses in petroleum and mining engineering at the Colorado School of Mines, Virginia Tech, South Dakota School of Mines and Technology, and Marietta College [6,14]. As a whole, our teaching reached over 1200 students. In our prior research [6], we found that students in all of the courses improved in defining CSR, especially in recognizing its intertwined social, environmental, and economic dimensions and in recognizing a broader array of stakeholders. Depending on the course, the majority (between 70–100%) ended the courses believing that CSR would be relevant to their careers as engineers, which potentially upsets the social/technical dualism that would define engineering as purely technical work and CSR as the responsibility of social scientists. We did not, however, find that students ended our courses expressing greater desires to work for companies with positive reputations for CSR, perhaps because they took a pragmatic view of their job market possibilities.

This article builds on that prior research by investigating whether and how our teaching shifted petroleum engineering students' understanding of the sociotechnical nature of both CSR and engineering. To assess the impact of our teaching enhancements on students' knowledge, attitudes, and skills, we developed and validated a survey instrument [14]. It included themes of corporate social responsibility, the ethical dimensions of engineering practice, engineers' agency in the workplace, students' career desires, and demographic information. In each course, all students took the survey at the beginning and end of the semester so that we could compare their responses before and after our course activities. We assigned each student who provided informed consent to participate in the research a unique and anonymous ID to match their pre- and post-course surveys (and track them year-by-year, for cohorts that participated in multiple classes) and calculated average responses for each class. We coded the qualitative responses. Finally, we collected end-of-semester reflections for the two senior level courses to generally assess student attitudes.

We focus on five classes of petroleum engineering students in two course offerings in this paper: three semesters of Summer Field Session I (summers 2017, 2018, and 2019), and two semesters of Senior Seminar (Fall 2016 and 2017). Both courses are required, meaning that they enroll the full cohort of petroleum engineering majors. The full demographic information for those courses can be found in [14], though we note that female students usually comprise a minority of the classes (30% and below) and there is a significant number of international students, especially from the Middle East (around 20%). The Summer Field Session enrolled students between their sophomore and junior year. The students traveled as a group and were introduced to the petroleum engineering industry through site visits, company tours, guest speakers, and facility tours. The Senior Seminar was designed to build their professional skills in the final year of their undergraduate program. The curriculum included approaching CSR through role-playing activities, a series of case studies based on actual experiences of an alumnus, industry speakers, and projects focused on controversial issues in the petroleum industry. For both of these course offerings, the intent was to promote interest, value, and motivation for learning about CSR; connect it to future careers; and integrate social and technical dimensions of petroleum engineering.

2.2. Sociotechnical Approaches to ASGM in Mining Engineering

The second area of research and teaching we analyze is the NSF-funded Responsible Mining, Resilient Communities (RMRC) project, which is an international and interdisciplinary effort to co-design socially responsible and sustainable gold mining practices with communities, engineers, and social scientists. The project focuses on artisanal- and small-scale gold mining (ASGM) in Colombia and Peru. While ASGM is an internally variegated field of practice, it generally refers to "labour-intensive, low-tech mineral exploration and processing" [15]. Most ASGM is done by individuals or small crews and happens without a title, making it an informal economic activity–and sometimes an illegal one–that exists in tension with government entities.

A key focus of the project is training undergraduate engineering students to approach ASGM from a sociotechnical perspective. In our research, we are investigating whether program activities enhance students' global sociotechnical competency. Building from prior research [16–18], we define global sociotechnical competency as being built from sociotechnical coordination; understanding and negotiating engineering and relevant national or local cultures; navigating ethics, standards, and regulations; and socially responsible engineering [19]. Table 1 provides an overview of the knowledge, skills, and attitudes that relate to each of these dimensions, using ASGM as an example.

Our prior research investigated whether participation in an intensive summer RMRC fieldwork session with Colombian faculty, students, and stakeholders enhanced U.S. undergraduate students' global sociotechnical competency. Because of the COVID-19 pandemic, we were able to test three different types of field sessions: one fully in person, in which US students traveled to Colombia (2019); one fully remote, in which students participated in activities on virtual platforms from their own work spaces (2020); and one hybrid, in which U.S. students studied together on a college campus but connected virtually with Colombian stakeholders (2021). We found that all three field sessions enhanced students' global sociotechnical competency. In particular, students ended the field sessions with a greater ability to identify the inherent social dimensions of problems that appear to be "technical" and with a greater ability to identify diverse stakeholders [19].

The current article builds on that prior research by investigating whether and how a week-long exchange at the Colorado School of Mines influenced how the Colombian students thought about the sociotechnical nature of ASGM. The delegation of visitors included twelve students and one faculty from the Universidad Nacional de Colombia's School of Mines in Medellín, plus one faculty from SENA's Centro Minero Ambiental in El Bagre (Colombia). They were hosted by RMRC faculty and students from Mines and the University of Texas at Arlington.

The delegation of Colombian students consisted of twelve students, ten women and two men, from the School of Mines of the Universidad Nacional de Colombia in Medellín. Eleven were from the Mining Engineering and Metallurgy program and one student came from the Environmental Engineering program. All the students had completed and passed more than 80% of the academic program and all of them were active members of the student chapter of SME (Society for Mining, Metallurgy, and Exploration). Most of them had participated in the joint work programs with the Colorado School of Mines in the two previous years. In their training they had a sociotechnical approach to mining projects in Colombia, primarily artisanal and small-scale gold mining. The students came from similar socioeconomic backgrounds. Some had relatives linked to the mining activity in Colombia, either as engineers or workers in a national mining company. One student came from an artisanal mining family. It is important to highlight that although all the students had a very good level of English, the exchange was the first opportunity for many of them to travel outside of the country.

Table 1. Global socio technical competency framework, originally published in [19].

Content Dimensions Learning Outcomes	Sociotechnical Coordination	Understanding and Negotiating Engineering and National Cultures	Navigating Ethics, Standards and Regulations	Socially Responsible Engineering
Knowledge	Understanding ASGM as a sociotechnical system	Understanding the history and political economy of ASGM in different countries Understanding the history and political economy of engineering in different countries with ASGM	Understanding legal dimensions of mining, labor & environmental management that affect ASGM	Understanding power differentials, how to have empathy, build trust, and treat expert and non-expert stakeholders involved in ASGM
Skills	Ability to identify different stakeholders in the ASGM life cycle and mediate among their needs and desires Ability to see how "technical" and "social" dimensions of ASGM co-constitute each other	Ability to operate differently in ASGM in different countries Ability to work with engineering faculty from different countries with ASGM	Ability to consult experts to ensure that sociotechnical innovations/design projects comply with legal and other regulatory standards relevant to ASGM	Ability to listen, engage in perspective taking, operate within different power positions, and work with expert and non-expert stakeholders involved in ASGM
Attitudes	Willingness to work with expert and non-expert stakeholders along the ASGM lifecycle Willingness to open up engineering decision making to a variety of social perspectives	Willingness to work with different ASGM perspectives in different countries and engineering faculty from different countries	Willingness to ensure that sociotechnical innovations/design projects comply with legal and other regulatory standards relevant to ASGM	Willingness and desire to engage in perspective taking Willingness and desire to work with expert and non-expert perspectives during project and after graduation Willingness and desire to use engineering to serve underprivileged populations Confidence in being able to make positive changes in communities through engineering

The exchange included a mix of field trips, lectures, and workshops. Participants took a field trip to a historic mining region in the Colorado mountains, where they were able to visit one of the world's largest molybdenum mines and the National Mining Museum and Hall of Fame. Students toured labs and centers at Mines, including a geoscience-themed makerspace, the Space Resources lab, the Earth Mechanics Institute, and the Geology Museum. They met and listened to presentations from engineering and social science faculty from the university's Humanitarian Engineering program, Payne Institute for Public Policy, and Instituto para Iniciativas Latino Americanas (Institute for Latin American Initiatives). They participated in workshops on asset-based community development, social innovation, and creative capacity building that were led by faculty and practitioners from MIT's D-Lab (whose mission is design for a more equitable world); Corps Africa (a non-profit that trains Africans in international development); and the Universidad Minuto de Dios (Colombia) Parque Científico de Innovación Social (Scientific Park for Social Innovation).

At the end of the exchange, students filled out a survey that included previously validated questions about their global socio technical competency.

3. Results

3.1. *Sociotechnical Learning in Petroleum Engineering*

We begin with a quantitative analysis of student responses to a survey question that asked students to evaluate CSR activities:

Q: CSR is a diverse field of practice that varies by industry, location, and company. In this survey we use an umbrella definition for CSR: an approach to business in which companies collaborate with stakeholders to create shared economic, social and environmental value. How would you evaluate the following activities as potential examples of CSR?

The possible responses ranged from primarily "social" activities (such as community training) to those that were sociotechnical and directly engaged engineering itself (such as rerouting a problematic pipeline). Students characterized each as being an excellent example of CSR, an okay example of CSR, or not CSR, with the option of selecting "I don't know". Of the possible responses, the three that most reflect a sociotechnical approach to engineering and CSR are underlined:

- A company providing training for members of a local community who want to open their own small businesses
- A team of engineers redesigning an industrial process to minimize potential spills of hazardous materials after learning that residents are worried about pollution
- A company giving college scholarships to children in the community where they operate
- A company accurately and transparently reporting how much money it spends in another country
- Employees doing charity or volunteer work in their free time
- A company constructing a municipal wastewater treatment plant for a city that desires but does not have one, so that the company can reuse the treated wastewater in its own production process
- An engineer reporting an unsafe practice to management or government authorities
- A company prioritizing local residents when making hires for new jobs
- An engineer changing the route of a pipeline to mitigate community conflict even though it will cost the company more money

Overall, the five courses were effective in helping students identify the three underlined "technical" decisions as also CSR decisions. In each course, more students ended the semester being able to identify at least two of the three sociotechnical CSR examples (the first two columns of Figure 1). In only one course (Fall 2017 Senior Seminar) did large numbers of students move away from "OK example" to either "excellent," "not CSR," or "I don't know." We explain potential reasons for this outcome below.

The full data set is available in Table 2 summarizes student assessments from five Petroleum Engineering classes for the three underlined options. We received unique responses from 427 students over the five classes.

Table 2 shows that the changes we observed in the student responses from the beginning to the end of the semester varied by both course and year. For example, fewer students in the Fall 2016 Senior Seminar ended the semester assessing redesigning an industrial process as excellent CSR (down to 70% from 81%), but more judged building a water treatment plant and rerouting a pipeline as excellent (up to 64% and 64% from 58% and 53%, respectively). The 2017 cohort showed improvements in judgements of excellent for each example. In the 2017 and 2019 summer field sessions, there were larger jumps in improvement for recognizing each example as excellent, but in 2018 fewer students judged building the treatment plant as excellent at the end.

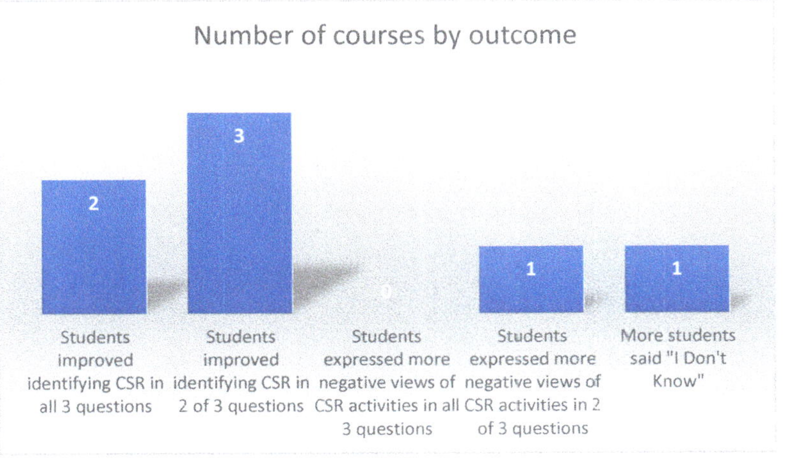

Figure 1. Course outcomes for Mines students (out of a total of 5 courses).

There was much more uncertainty indicated for the 2017 Senior Seminar cohort across all three questions, as indicated in the increase from pre- to post-survey responses of "not CSR" and "I don't know." Growing awareness of the complexities of practicing engineering, along with the beginning of a serious downturn in the petroleum industry may have led to more polarization in the students' views of CSR. During this time, several students' job offers were rescinded and oil and gas companies had more and more challenges keeping their doors open, and less money to incorporate multiple stakeholders' needs. This polarization is illustrated well with this student end-of-semester reflection on the seminar course:

Personally, I think this class is very interesting and perhaps my favorite, contrary to most of the people I've asked. I feel it is super important to broaden our horizon into the non-technical aspects of the industry, especially for those who would like to be leaders in the industry and make a positive impact. However, some/most of my colleagues think otherwise .... I believe that a lot of the student's frustrations with the course are tied to lack of opportunities in the industry, and the fact that this course "steals" time for other studying to be conducted. Especially in a time where students are trying to boost their GPA, with the belief that it is their best method to increase chances of employment.

There was more uncertainty and skepticism present in each of the senior seminars, which is likely due to the timing of the courses in the students' undergraduate progression. The field session is taken by students just entering the petroleum engineering major, while the seminar is taken by students who are typically in their last year of the program. This difference may have led to more skeptical evaluations of potential CSR activities by the senior students, as many of them would have had much more exposure to technical topics along with possible internships related to the petroleum industry. Thus their views are much more sophisticated, technical, and prone to influence from companies and current events. With increased technical knowledge, yet limited broad industrial knowledge, for example, seniors may only interpret "redesigning industrial processes" as a technical intervention, rather than making the connections to the ways these changes may serve the public. Additionally, the seminar course was offered during the most intensive semester of petroleum engineering courses for these students. This led to many not taking it very seriously. One student commented that offering it a semester later could lead to more students valuing the course material and taking it personally: " I believe a relatively reduced course load in the Spring, combined with the fact that (some) students will finally realize that "the end of academics is near", will provide a sobering feeling that they need to broaden horizon to learn more about 'what's out there".

Table 2. Student assessments of CSR activities.

		Redesigning Industrial Processes		Building Treatment Plant		Rerouting Pipeline	
		Pre	Post	Pre	Post	Pre	Post
Senior Seminar Fall 2016	Excellent Example	81.08%	69.90%	58.11%	63.59%	52.74%	63.78%
	OK Example	13.51%	19.90%	25.68%	25.13%	30.14%	27.55%
	Not CSR	4.73%	8.67%	13.51%	9.23%	13.01%	4.59%
	I don't know	0.68%	1.53%	2.70%	2.05%	4.11%	4.08%
	Total students	148	196	148	195	146	196
Senior Seminar Fall 2017	Excellent Example	82.05%	79.49%	61.54%	74.36%	76.92%	80.77%
	OK Example	16.67%	7.69%	33.33%	12.82%	17.95%	12.82%
	Not CSR	1.28%	8.97%	5.13%	7.69%	3.85%	3.85%
	I don't know	0.00%	3.85%	0.00%	5.13%	1.28%	2.56%
	Total students	78	78	78	78	78	78
Summer Field Session 2017	Excellent Example	61.54%	74.36%	61.54%	76.92%	51.28%	71.79%
	OK Example	23.08%	20.51%	28.21%	20.51%	30.77%	23.08%
	Not CSR	15.38%	5.13%	7.69%	2.56%	12.82%	2.56%
	I don't know	0.00%	0.00%	2.56%	0.00%	5.13%	2.56%
	Total students	39	39	39	39	39	39
Summer Field Session 2018	Excellent Example	73.77%	83.61%	50.82%	45.90%	55.74%	73.33%
	OK Example	19.67%	13.11%	27.87%	36.07%	29.51%	18.33%
	Not CSR	6.56%	3.28%	16.39%	14.75%	9.84%	5.00%
	I don't know	0.00%	0.00%	4.92%	3.28%	4.92%	3.33%
	Total students	61	61	61	61	61	60
Summer Field Session 2019	Excellent Example	77.78%	88.89%	53.70%	64.81%	62.96%	70.37%
	OK Example	14.81%	7.41%	29.63%	27.78%	27.78%	25.93%
	Not CSR	5.56%	1.85%	14.81%	7.41%	7.41%	1.85%
	I don't know	1.85%	1.85%	1.85%	0.00%	1.85%	1.85%
	Total students	54	54	54	54	54	54

There were also several end-of-semester reflections from students about the difficulty of balancing the needs and desires of so many groups with different aims, which also point to an increasing sophistication in perception of how CSR plays a role in the work they hoped to do. This is summed up well with this student perspective: "The one thing I struggle with is finding a balance between the business side of myself, and the empathetic side of myself. The business side can easily come up with the key stakeholders that need to be addressed, but often overlooks the fact that the people with no voice and no one to protect them, desperately need advocates within the oil and gas industry to make sure they are not overrun. On the other hand, the empathetic side of me could happily take forever, and come up with a solution which perfectly fits everyone's needs-even though a perfect solution that makes everyone happy usually doesn't exist".

Student reflections and general instructor observations also provide some insight into how students' thinking shifted regarding CSR, socio-technical thinking, and reconciling

CSR and its complexities into their professional practice. Many students originally perceived CSR to be about environmental stewardship but came to appreciate that CSR also included social dimensions. Initially, students justified CSR as a way to make profits, share wealth, satisfy shareholders, and create jobs. That view became more nuanced as they also learned the importance of protecting reputation, mitigating risk, and maintaining a social license to operate. They shifted from viewing CSR as a way to promote the public good in general, to CSR as a specific way to implement sustainable community development and improve local quality of life. Concurrently, there was a noted shift from defining CSR as sharing benefits and being philanthropic to CSR better aligning with the Auld et al. definition of "new" CSR or redesigning core business practices. Student reflections indicated that many believed CSR to encompass more complex social responsibilities such as maintaining transparency and seeking mutual understanding. All of this was undergirded by a shift from examining problems as technical challenges to sociotechnical problems. Students observed that issues many stakeholders face regarding petroleum engineering projects are more-than-technical, and thus, they needed to find more-than-technical ways to address concerns. They also noted that many people had major concerns regarding the petroleum industry and that people want to be heard and understood, rather than being "assaulted" by facts. This suggests that they were able to see the "problem" of petroleum engineering depended on who was defining it, and that stakeholders could define the problem differently than a petroleum engineer.

3.2. Sociotechnical Learning in Mining Engineering

All students who participated in the exchange completed the survey and wrote brief reflections at the end of the session. Table 3 summarizes the average student responses to the survey questions related to global sociotechnical competency, which focused on working in unfamiliar places, collaborating with people from different backgrounds, empathizing, and feeling confident in being an engineer.

We found that students ended the exchange expressing strong desires to work and live abroad (5.0 out of 5.0) and serve underprivileged populations (4.9) Importantly, they also expressed confidence in working with engineering students from different backgrounds (4.8) and learning from professors with different backgrounds (4.9). Empathy is a crucial dimension of global sociotechnical competency, and students expressed comfort and enjoyment learning about unfamiliar people and places (4.6), talking with people from different backgrounds (4.7), asking people questions about their experiences (4.4), and seeing other people's point of view (4.6). They also expressed strong self-efficacy, including confidence in their abilities as engineers (4.4), and positive views of engineering as a fulfilling profession (4.5) that makes it possible to make positive changes in com-munities (5.0).

The survey also explicitly asked students about whether the visit provided them "new perspectives on engineering as a sociotechnical activity" and helped them "understand the social, environmental, and economic dimensions of mining." Students responded positively to both questions, with average responses of 4.9 out of 5 for the former and 4.8 out of 5 for the latter.

It seems likely that this more holistic view of engineering in general and ASGM in particular is related to the overwhelming sense that the exchange provided professional growth opportunities. In their written comments, students described the exchange as being a "mind changer" that gave them the "opportunity to see new perspectives in the mining industry," and as an experience that opened their eyes to "new possibilities" for their professional careers. Many of them referenced the sociotechnical theme of week–and seeing how they can contribute to ethical goals through their professional practice–as being transformative. The following are quotes from students:

- "Now, I understand that there must be a balance between many aspects such as: ethical, humanitarian and environmental."
- "It is a mind change to become a person that contributes to community development from science."

- "I had a huge desire to contribute to science but [now I know] that I want to con-tribute to science but also serve underprivileged communities."
- "The most valuable aspect to me was being able to integrate all the social, environmental and technical aspects of mining engineering. It was a very enriching experience that would allow me to continue improving as a professional and a person."
- "This visit reinforced my ideals of combining social knowledge with technical knowledge and I was able to make many contacts with excellent professors from different universities."
- "This visit allowed me to open my mind to more possibilities in the mining sector that I didn't know so far. I was able to discover how topics I have always been passionate about can have applications in mining."

Table 3. Average student self-assessments on a scale of 1 to 5 (1 = not at all like me; 3 = neutral; 5 = very much like me).

Question (1 Is Low, 5 Is High)	Average
I like to learn about people and places unfamiliar to me.	4.6
I feel comfortable talking with people from different backgrounds.	4.7
I like to ask people questions about their experiences.	4.4
It is easy for me to see other people's points of view.	4.6
I feel confident working with engineering students from different backgrounds.	4.8
I enjoy learning from professors from different backgrounds.	4.9
I would like to study or work internationally at some point in my career.	5.0
I would like a career that allows me to serve underprivileged populations.	4.9
I am confident in my abilities as an engineer.	4.4
I find fulfillment in engineering.	4.5
I can make positive changes in communities through engineering.	5.0

After this experience, the students reflected on how they had broadened their knowledge of new ways of learning. They developed a greater tolerance to work in difficult conditions and an approach to other methodologies of interaction with the environment, both large-scale, which some of them already knew, and small-scale, which represented a novelty for others. All of them highlight this experience as very positive and formative and appreciate the sustainability of mining as a central axis of their professional performance.

4. Discussion

Both sets of students–the Mines petroleum engineering students and the Colombian mining engineering students–ended their experiences with a greater knowledge of the sociotechnical nature of their chosen professions. The significant differences between the students, their experiences, and the assessments guard against tight comparisons. For example, almost all of the Colombian students were much more energized by the experience of coming to view mining engineering as a sociotechnical activity and described it as a formative moment in their professional development. We noted similar excitement in a portion of the petroleum engineering students, but that was tempered by resistance to the course material and activities among others. In a sense, more petroleum engineering students articulated more strongly that the "social" material was external to their core identity and responsibilities as engineers. This difference could be attributed to the different paths that led to the students participating in the sociotechnical learning opportunities: the Colombian students had all volunteered for the exchange and were not being graded on their performance, whereas the Mines students were required to take the courses for grades. Grades took on an added significance when the petroleum market downturn made competition for jobs fierce.

While one of the interesting findings among the Mines students was the increased polarization of opinion and uncertainty by their senior year, our instruments for the Colombian exchange did not allow us to measure uncertainty and polarization. We do note relative uniformity in the students' answers to the survey question: almost all students responded to questions with either a 4 or 5 on a 5-point scale, with 5 representing the most positive answer. It could be that the students were eager to show their appreciation for the trip, and so answered the questions extra positively.

The comparison of the student groups seems to point to the significance of real-world experiences as transformative for students' learning. Both the petroleum engineering field session and the Colombian mining exchange included visits to industrial sites and interactions with industry professionals, in addition to learning from their professors. In the senior seminars, the professor created multiple opportunities for industry connections–such as through invited guest speakers–but most of the activities took place in the classroom. These observations underscore our previous research that also showed that connections with practicing engineers was especially transformative for student learning about social responsibility [6].

Finally, we underline that all of the petroleum engineering courses and the Colombian exchange were the result of interdisciplinary collaborations among faculty from engineering and social science backgrounds. These kinds of collaborations are particularly well-suited to sociotechnical teaching and learning [3,8].

5. Conclusions

Some of the greatest contemporary challenges facing the mining and petroleum industries are sociotechnical in nature, dealing with thorny issues of public acceptance and social and environmental justice. Training the next generation of engineering students to approach problems from a sociotechnical perspective is a key strategy for addressing those challenges and developing industry projects that are responsive to local concerns and needs. Undergraduate education is a time in which students are not just developing technical expertise, but their own identities as engineers. Presenting students with social content directly inside of their majors is a powerful strategy for defining societal concerns as central to their responsibilities as engineers. Our teaching and research with two different groups of students–petroleum engineering students enrolled at the Colorado School of Mines (though hailing from around the U.S. and the world) and Colombian mining, metallurgical, and environmental engineering students from their School of Mines–found that collaborative, interdisciplinary teaching about authentic problems enhanced students' abilities to understand their professions from a sociotechnical perspective. This recognition is a crucial step to being able to then practice engineering in a way that promotes sustainability and social justice.

Author Contributions: J.S.: Conceptualization; investigation; methodology; formal analysis; resources; data curation; writing—original draft preparation; writing—review and editing; supervision; funding acquisition; project administration. C.M.: Investigation; formal analysis; writing—original draft preparation; writing—review and editing. O.J.R.: Supervision; writing—original draft preparation; writing—review and editing; supervision; funding acquisition. All authors have read and agreed to the published version of the manuscript.

Funding: This material is based upon work supported by the National Science Foundation under Grant No. (grantee must enter NSF grant 1743749. Any opinions, findings, and conclusions or recommendations expressed in this material are those of the author(s) and do not necessarily reflect the views of the National Science Foundation.

Data Availability Statement: Due to requirements for confidentiality for human subjects research, we have not made our data publicly available.

Acknowledgments: We thank Juan Lucena and Cristin Georgis for their collaboration in hosting the international exchange and Kate Smits for her participation. We also thank Linda Battalora for her role in integrating CSR into her courses in petroleum engineering, and Joey Tucker for his enthusiastic contribution to our petroleum engineering CSR case study.

Conflicts of Interest: The authors declare no conflict of interest.

References

1. Bijker, W.E. *Of Bicycles, Bakelites, and Bulbs: Toward a Theory of Sociotechnical Change*; MIT Press: Cambridge, MA, USA, 1997.
2. Johnson, D.G.; Wetmore, J.M. *Technology and Society: Building our Sociotechnical Future*; MIT Press: Cambridge, MA, USA, 2008.
3. Leydens, J.A.; Johnson, K.E.; Moskal, B.M. Engineering student perceptions of social justice in a feedback control systems course. *J. Eng. Educ.* **2021**, *110*, 718–749. [CrossRef]
4. Owen, J.R.; Kemp, D.; Lechner, A.M.; Harris, J.; Zhang, R.; Lèbre, É. Energy transition minerals and their intersection with land-connected peoples. *Nat. Sustain.* **2022**, *6*, 203–211. [CrossRef]
5. Owen, J.R.; Kemp, D.; Harris, J.; Lechner, A.M.; Lèbre, É. Fast track to failure? *Energy transition minerals and the future of consultation and consent. Energy Res. Soc. Sci.* **2022**, *89*, 102665. [CrossRef]
6. Smith, J.M. *Extracting Accountability: Engineers and Corporate Social Responsibility*; MIT Press: Cambridge, MA, USA, 2021.
7. Smith, N.M.; Zhu, Q.; Smith, J.M.; Mitcham, C. Enhancing Engineering Ethics: Role Ethics and Corporate Social Responsibility. *Sci. Eng. Ethics* **2021**, *27*, 28. [CrossRef] [PubMed]
8. Leydens, J.A.; Lucena, J.C. *Engineering Justice: Transforming Engineering Education and Practice*; Wiley-IEEE Press: Piscataway, NJ, USA, 2017. [CrossRef]
9. Faulkner, W. 'Nuts and Bolts and People': Gender-Troubled Engineering Identities. *Soc. Stud. Sci.* **2007**, *37*, 331–356. [CrossRef]
10. Cech, E.A. Culture of Disengagement in Engineering Education? *Sci. Technol. Hum. Values* **2014**, *39*, 42–72. [CrossRef]
11. Cohen, B.; Rossmann, J.S.; Bernhardt, K.L.S. Introducing Engineering as a Socio-technical Process. In Proceedings of the ASEE Annual Conference & Exposition, Indianapolis, IN, USA, 15–18 June 2014. [CrossRef]
12. Hoople, G.D.; Choi-Fitzpatrick, A. *Drones for Good: How to Bring Sociotechnical Thinking into the Classroom*; Synthesis Lectures on Engineers, Technology, & Society; Springer: Berlin/Heidelberg, Germany, 2020; Volume 9, p. i-148. [CrossRef]
13. Auld, G.; Bernstein, S.; Cashore, B. The New Corporate Social Responsibility. *Annu. Rev. Envir. Resour.* **2008**, *33*, 413–435. [CrossRef]
14. Smith, J.M.; Smith, N.M.; Rulifson, G.; McClelland, C.J.; Battalora, L.A.; Sarver, E.A.; Kaunda, R.B. Student Learning about Engineering and Corporate Social Responsibility: A Comparison Across Engineering and Liberal Arts Courses. ASEE Annual Conference & Exposition 2018. Available online: https://peer.asee.org/student-learning-about-engineering-and-corporate-social-responsibility-a-comparison-across-engineering-and-liberal-arts-courses (accessed on 1 June 2023).
15. Hilson, G. Artisanal mining, smallholder farming and livelihood diversification in rural sub-Saharan Africa: An introduction. *J. Int. Dev.* **2011**, *23*, 1031–1041. [CrossRef]
16. Downey, G.L.; Lucena, J.C.; Moskal, B.M.; Parkhurst, R.; Bigley, T.; Hays, C.; Jesiek, B.K.; Kelly, L.; Miller, J.; Ruff, S.; et al. The Globally Competent Engineer: Working Effectively with People Who Define Problems Differently. *J. Eng. Educ.* **2006**, *95*, 107–122. [CrossRef]
17. Jesiek, B.K.; Zhu, Q.; Woo, S.; Thompson, J.; Mazzurco, A. Global Engineering Competency in Context: Situations and Behaviors. *Online J. Global Eng. Educ.* **2014**, *8*, 1–14.
18. Smith, J.M.; Lucena, J.C. Socially Responsible Engineering. In *Routledge Handbook of the Philosophy of Engineering*; Michelfelder, D., Doorn, N., Eds.; Routledge: New York, NY, USA, 2021; pp. 661–673.
19. Smith, J.M.; Lucena, J.C.; Rivera, A.; Phelan, T.; Smits, K.; Bullock, R. Developing Global Sociotechnical Competency Through Humanitarian Engineering: A Comparison of In-Person and Virtual International Project Experiences. *J. Inter. Eng. Educ.* **2021**, *3*, 5. Available online: https://digitalcommons.uri.edu/jiee/vol3/iss1/5/ (accessed on 1 June 2023).

mining MDPI

Article

Toward Automatic Monitoring for Anomaly Detection in Open-Pit Phosphate Mines Using Artificial Vision: A Case Study of the Screening Unit

Laila El Hiouile [1,2,3,*], Ahmed Errami [1] and Nawfel Azami [3,4]

1 Engineering Research Laboratory (LRI), Networks Embedded Systems and Telecommunications Team (NEST), National and High School of Electricity and Mechanic (ENSEM), Hassan II University of Casablanca, 5366 Maarif, Casablanca 8118, Morocco; a.errami@ensem.ac.ma
2 Research Foundation for Development and Innovation in Science and Engineering (FRDISI), Casablanca 20100, Morocco
3 Analytics Laboratory, Mohammed VI Polytechnic University (UM6P), Ben Guerir 43150, Morocco; azami@inpt.ac.ma
4 Optics Laboratory, National Institute of Posts and Telecommunications (INPT), Av. Allal Al Fassi, Madinat Al Irfane, Rabat 10000, Morocco
* Correspondence: leila.elhiouile@ensem.ac.ma

Abstract: Phosphorus is a limited resource that is non-replaceable worldwide. Its significant role as a fertilizer underlines the necessity for prudent and strategic management. The adequate monitoring of the phosphate extraction process mitigates anything that can influence the quantity or quality of the product. The phosphate extraction process's most important phase is the screening unit, which can be used to separate phosphate minerals from unwanted materials. Nevertheless, it encounters several anomalies and malfunctions that influence the performance of the whole chain. This unit requires continuous automated control to avoid any blockages or risks caused by malfunctions. Using artificial intelligence and image processing techniques, the main goal of the investigations described in this paper was to evaluate the performances of machine-learning and deep-learning models to detect the screening unit malfunction in the open pit of the phosphate mine in Benguerir-Morocco. These findings highlight that the CNN and HOG-based models are the most suitable and accurate for the given case study.

Keywords: open-pit phosphate mine; phosphate ore screening unit; anomaly detection; intelligent monitoring system; machine learning; deep learning

Citation: El Hiouile, L.; Errami, A.; Azami, N. Toward Automatic Monitoring for Anomaly Detection in Open-Pit Phosphate Mines Using Artificial Vision: A Case Study of the Screening Unit. *Mining* **2023**, *3*, 645–658. https://doi.org/10.3390/mining3040035

Academic Editor: Juan M. Menéndez-Aguado

Received: 16 September 2023
Revised: 16 October 2023
Accepted: 18 October 2023
Published: 20 October 2023

1. Introduction

The phosphate industry is one of the most essential industries in the world since phosphorus is an irreplaceable resource in agriculture and, at the same time, it is limited in terms of availability [1]. The phosphorus fertilizer is in high demand for crop production [2–4] because its use can improve yields by up to 50% [5]. Phosphorus fertilizers consume more than 80% of the phosphorus produced. Otherwise, phosphorus and its compounds are used in animal feed, detergents, and operations for metal processing [2]. Therefore, this resource must be managed adequately to prevent or at least minimize future supply limitations. A part of this stewardship plan is to perform quality control analyses through real-time monitoring to ensure the final product's quality and improve extraction yields.

Manual monitoring is more than monotone and leads to errors that are overlooked by humans; it is also often impractical, even as production volume increases, not to mention costly. In fact, the surveillance officer can only maintain an acceptable level of attention for up to 20 min when observing and analyzing video surveillance monitors and can keep an attentive eye on 9 to 12 cameras for up to 15 min [6]. Thus, intelligent video surveillance systems could provide a solution to the limitations of manual human monitoring. This kind of intelligent system is one of the primary objectives of the Fourth Industrial Revolution.

Thus, constructing intelligent plants to modernize manufacturing processes is key to innovation, growth, and sustainable profitability. Several works have explored various aspects of intelligent monitoring. For example, in our specific context, work [7] presents a complete system that was designed to facilitate the condition monitoring of railway tunnels by structural examiners. This technology increases accuracy and robustness while reducing the time required for visual inspection. On the other hand, ref. [8] illustrates an example of intelligent video surveillance that was designed to automatically detect hex head bolts used to fasten rails to sleepers. This system is based on MLPNC (Multi-Layer Perceptron Neural Classifier) and FPGA (Field-Programmable Gate Array) technologies. Ref. [9] classifies the current mining applications of UAVs (Unmanned Aerial Vehicles) from exploration to reclamation. At the same time, video surveillance has been used in the mining context to anticipate risks and improve mining safety and productivity, as shown in the work [10]. The latter proposes a hybrid CNN-LSTM (convolutional neural networks and long short-term memory networks) prediction model to accurately anticipate miners' health quality index and CH4 gas concentration. Finally, paper [11] presents a model to automatically identify and monitor open-pit mines in Hubei province, China, by exploiting Gaofen-2 and Google Earth satellite data using the R-CNN (region convolutional neural network) and transfer learning. These works contribute substantially to the progress of intelligent monitoring in different areas. However, notable gaps remain in the intelligent monitoring of the phosphate production chain associated with its unique challenges. Our research seeks to fill these gaps and provide insights into the specific challenges of the phosphate screening unit, which fundamentally influences the entire production process. This monitoring offers multiple benefits by automating the control of the screening unit and detecting anomalies, thereby considerably improving the yield of the phosphate production chain. Simultaneously, it significantly reduces machine maintenance costs, representing a significant financial advantage for mining operations.

Thus, we aim to provide a reliable and effective video surveillance system that can detect malfunctions in the Benguerir phosphate mine screening unit using computer vision and artificial intelligence tools. In previous works [12,13], we demonstrated that certain models provide enhanced results in the classification of anomalies within the screening unit. However, in this work, we extended our investigation by evaluating additional techniques known for their robustness in anomaly classifications. Our aim was to identify the most effective models capable of maintaining their performance in the presence of various future perturbations. The rest of this paper is organized as follows: Section 2 illustrates phosphate industry malfunctions/anomalies in the Benguerir mining site and their consequences, explaining the need for intelligent solutions. Section 3 explains the methods and materials. Section 4 presents the implementation process with the obtained results. Finally, we provide a discussion and then a conclusion.

2. Malfunctions of the Phosphate Production Chain in the Benguerir Mining Site

Morocco holds three-quarters of the world's phosphate reserves, making it the world's leading exporter with around a 1/3 of international trade, the world's leading exporter of phosphoric acid (50% of the international market), and the world's third largest phosphate producer. Despite its economic importance and beneficial effects, this status represents a major responsibility and a real challenge regarding the safeguarding of this resource against any loss or damage.

Malfunctions in any production process mean product loss, which negatively impacts the production line's yield. In the phosphate industry, malfunctions and phosphate losses really affect ore recovery rates. There are two main sources of phosphate losses during mining [14]:

- Project losses: there are losses of phosphate in places that have been abandoned and not mined; they involve the abandonment of phosphate levels whose mining generates very high ratios and is, therefore, economically unfeasible.
- On-site losses: there are losses linked to different operational stages, from the kinematic chain that extracts the various phosphate layers to the final loading of the product.

At the Benguerir phosphate mine, operations begin with extracting phosphate layers, which consists of the following stages: drilling, blasting, stripping, and phosphate ore recovery. Once the ground has been drilled, explosives are placed in the holes and then blasted to reshape the ground and make it crumbly for easy stripping. Stripping is the operation that consists of removing the overburden or intervening layers to expose and recover the layer of ore to be exploited. Next, the ore (phosphate) is recovered once stripping is complete. The ore then undergoes a series of treatments, mainly destoning and screening, to reduce the quantity of waste rock and ensure the required product's quality.

Ref. [14] presents the problems and losses encountered during phosphate ore recovery at the Benguerir mining site in Morocco. This paper highlights problems and losses in relation to staff qualifications and lack of supervision, as well as other challenges linked to the soil's nature concerning the adaptation of equipment used, the encumbrance of impurities on the phosphate, and problems related to drilling, blasting, cleaning, and transport operations. Following the phosphate ore recovery operation, the phosphate ore beneficiation process begins. Effective beneficiation can be achieved through various processes, depending on the liberation size of phosphate, gangue minerals, and other ore specifications [13]. The screening operation is one of the most effective beneficiation processes used in the Benguerir mine. The screening station contains a certain number of screens, which are used to separate the phosphate minerals from unwanted materials. As part of this research project, we had the opportunity to visit the Benguerir site and received a detailed report on the various anomalies encountered at the phosphate screening station. This station has an intrinsic role, meaning that any dysfunction during this stage directly affects the overall effectiveness of the process.

The main problems encountered in the screening unit can be generalized into two primary anomalies: the abnormal presence of sterile stones on screens, which negatively impacts the quality of the final product, and the rejection of high-quality phosphate. Indeed, the machines cannot eliminate the stones mixed with the phosphate during the de-stoning operation. As a result, the stones that cannot be removed could block the hoppers' opening in the main screening building and create a blockage in the production line that can last anywhere from half an hour to eight hours. The situation worsens when a poor phosphate ore layer is extracted. The screens overflow with waste rock contained in low-concentration phosphate ore. The screens may be unable to re-screen the product because the mesh is blocked. The existence of large quantities of phosphate mixed with waste rock is another malfunction that leads to the loss of large quantities of net product due to a delay in detecting the root cause of the problem. In some cases, the screens cannot filter all the material due to the high flow rate of the material. Figure 1 presents images illustrating malfunctions occurring in the screening unit at the Benguerir site.

These malfunctions affect the screening process in several ways, with infiltrated sterile stones producing a direct negative impact on (i) safety, (ii) production yield due to machine stoppages and micro-stoppages, (iii) machine life due to the vibrations produced by large stones, which have an impact on maintenance costs, and (iv) the loss of production caused by the passage of material to screen rejection.

Generally, the gravity of all these malfunctions lies in a delay in detection, resulting in ineffective intervention by the maintenance department. Therefore, the screening operation must undergo quality control analyses via real-time monitoring using surveillance cameras and intelligent computer vision and machine-learning techniques to automate surveillance and anomaly detection.

Figure 1. Illustration of some malfunctions in the screening unit at the Benguerir site. (a) Passage of pebbles to screens on the conveyor belt. (b) Pebbles in the screen. (c) High sterile content in the screen. (d) Passage of the product through the screen.

3. Materials and Methods

3.1. Method

The choice of monitoring method is mainly based on the information available in the system. Empirical feedback is represented by system expertise, historical data recorded after using the system under various conditions, physical models derived from a basic understanding of the system, and the physics of the system, expressed as a mathematical function in relational form [15]. There are two categories of monitoring approaches: model-based and data-based.

Model-based detection and diagnosis offer a description of dynamic behavior and a better physical understanding of the system, which is a major advantage. However, in practice, it is very difficult to develop an accurate mathematical model that considers the different sources of uncertainty due to the complexity of systems. The model-based approach is generally applied on the assumption that only simple failures occur. However, when a large amount of historical data are available, data-driven approaches are a good alternative [16].

Most methods based on historical data consider detection and diagnosis as classification tasks (see Figure 2). The aim of detection is to identify whether an abnormal operation has occurred, which corresponds to a classification into two categories: the normal functioning class (NFC) and the fault class. Diagnosis aims to determine the type of fault, which can be seen as a classification into several classes: anomaly 1, anomaly 2, etc. [15].

In this work, we adopted a data-driven approach and investigated the classification technique based on supervised learning. We considered two main anomalies of the screening unit:

- Anomaly 1: High sterilization rate.
- Anomaly 2: The passage of phosphate material to screen rejection (phosphate loss).

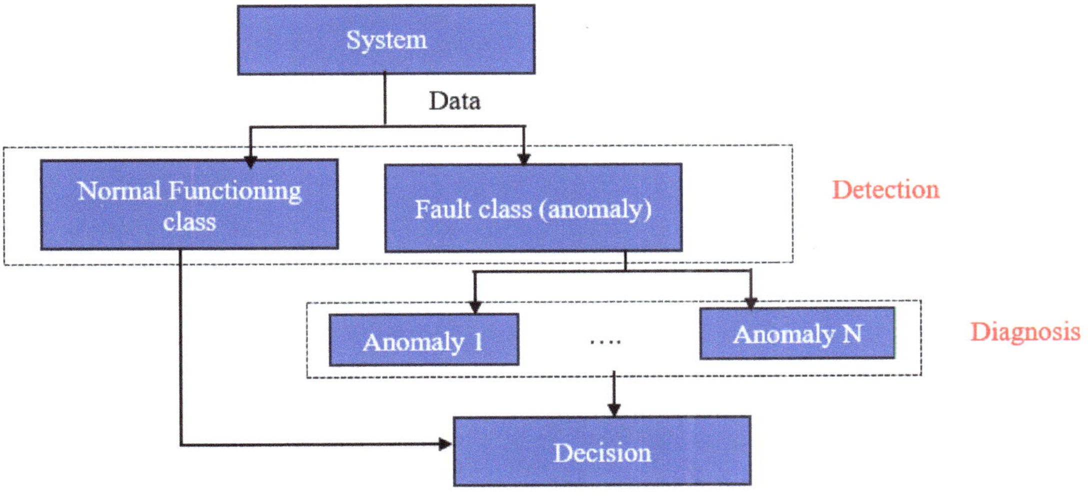

Figure 2. Detection and diagnosis of anomalies based on a data-driven approach.

Therefore, we approached anomaly detection as a problem of image classification into three distinct classes: the NFC class, anomaly 1 class, and anomaly 2 class. Figure 3 illustrates our implemented method for detecting malfunctions using image classification with three classes.

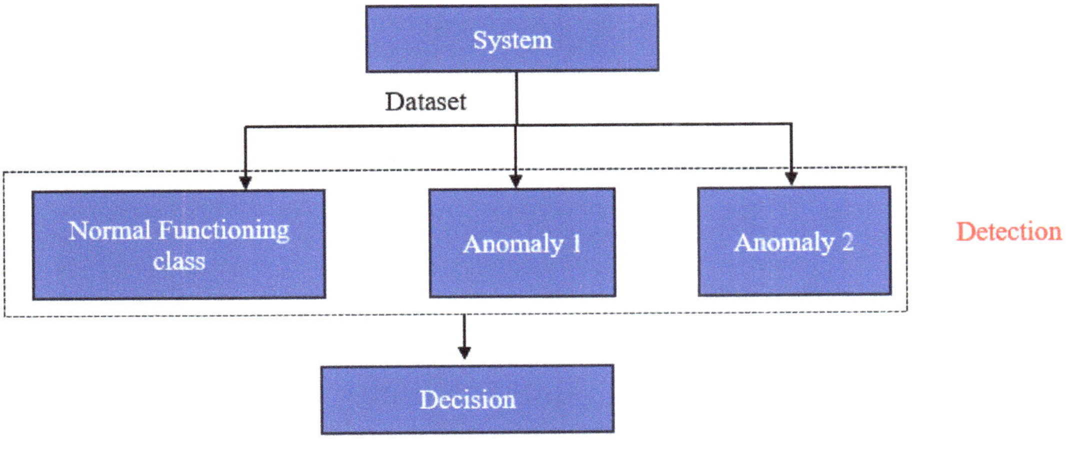

Figure 3. Our method based on a data-driven approach and supervised learning-based classification technique.

Various methods have been developed for image classification tasks. There are mainly traditional machine learning techniques such as support vector machines (SVM), k-nearest neighbors (KNN), and random forest (RF), as well as deep learning techniques such as convolutional neural networks (CNN). Traditional machine learning techniques rely on hand-crafted features extracted from images using a feature extractor such as Histogram of Oriented Gradient (HOG) or Local Binary Pattern (LBP), while deep techniques automatically extract features using their convolutional layers. Although deep models have been the most used recently in many works, including industrial damage detection with excellent results such as [17–19], we believe that each problem has its own challenges. Hence, in this study, we conducted a comparative evaluation of machine-learning and deep-learning approaches to select the optimal models for our case study. For the machine-learning approach, we evaluated a combination of HOG, Scale Invariant Feature (SIFT), and LBP with one of the classifiers, SVM, KNN, or RF, while for the deep approach, we tested the CNN model.

The support vector machine (SVM) is a well-known classification algorithm. It seeks to create an optimal hyperplane, maximizing the separation between projected points, called support vectors. SVMs are versatile, handling both linear and non-linear classifications using kernel functions [20]. On the other hand, random forests, an ensemble learning method, build several decision trees during training. Each tree contributes a unit vote to classify an input vector based on the most common class [21]. K-nearest neighbors (KNN) is a simple and efficient non-parametric classification method that determines the class of a new data point by examining the majority class among its k-nearest neighbors from a set of labeled training data [22].

Furthermore, the convolutional neural network (CNN), a well-known model of feed-forward neural networks, is particularly well suited to large datasets such as images and videos. CNNs work the same way as standard neural networks, except that each unit of a CNN layer is a two-dimensional convolution filter applied to the layer's input. This convolution step is essential when learning models from high-dimensional inputs, such as images or videos [23]. Regarding feature extraction algorithms, Table 1 briefly describes the HOG, SIFT, and LBP algorithms, with an illustration of their application in an image corresponding to the situation with the high sterilization rate.

These techniques have proven their efficacy in various smart surveillance applications. For instance, the HOG descriptor has demonstrated its performance in human detection, tracking, and object detection [24,25]. LBP and Violent Flows (ViF), followed by Linear SVM, have been used to classify videos as either violent or non-violent [26]. Additionally, SIFT has proven its efficiency when used as a feature extractor for anomaly detection [27]. On the other hand, the widely used classification algorithm SVM has been employed, for example, when detecting abnormal events in public surveillance systems [28]. Random forest has been utilized to automate defect detection in tunnel images [29]. Moreover, the HOG-SVM combination has demonstrated its effectiveness in detecting anomalies in the screening unit [12]. However, this work aims to evaluate and compare other combinations to identify the most powerful models based on various evaluation metrics.

Figure 4 shows the flowchart used to build the models. Once the images were acquired, our knowledge database was built to train the different models evaluated, whether from classical or deep approaches. Finally, a comparative analysis was developed based on the evaluation metrics.

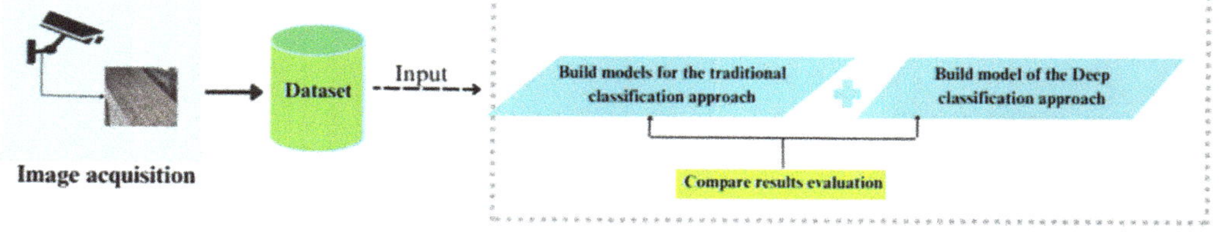

Figure 4. Flowchart of the methodology followed.

Table 1. HOG, SIFT and LBP principals and their application in an image corresponding to the situation of a high sterilization rate.

Algorithm	Principle	Application
HOG: Histogram of Oriented Gradient	HOG is a feature descriptor proposed by Navneet Dalal and Bill Triggs in 2005 [30] and used in computer vision for object detection. The basic principle of this descriptor is the use of the intensity distribution of the gradient or the direction of the contours.	 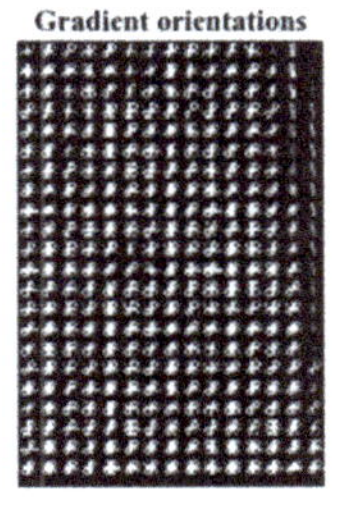
SIFT: Scale Invariant Feature Transform	SIFT is a feature extractor proposed by researcher David Lowe in [31]. The general idea of this algorithm is to extract characteristic points, called "features points", on an image in such a way that these points are invariant to several transformations, including rotation, illumination, and, especially, invariant to scale.	 
LBP: Local Binary Pattern	This descriptor was first mentioned in 1993 to measure an image's local contrast but was popularised three years later by Ojala et al. to analyze textures [32]; it is also used to detect and track moving objects in an image sequence. The general principle is to compare a pixel's luminance level with its neighbors' levels.	 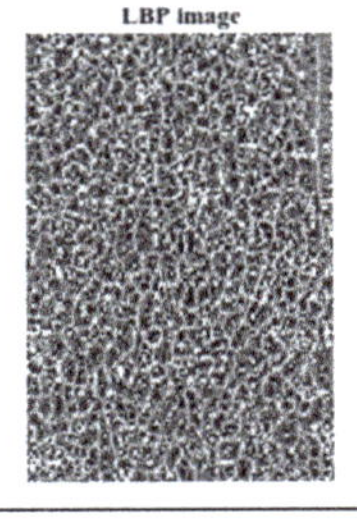

3.2. Datasets Preparation and System Configuration

The experiments were carried out on a balanced dataset containing images in the jpg format, each measuring 180 × 120 × 3, and captured from the videos of the surveillance camera installed at the screening station. The captured images were converted to grayscale images and then pre-processed to prepare a dataset of learning. For each captured image, a 32° rotation, a cropping, and a resizing operation were introduced to eliminate the non-functional parts of the image. This dataset contains three different classes; one class is the normal case, and the others present two types of anomalies (see Figure 5). Figure 5 shows images corresponding to the normal functioning class, images corresponding to the high sterilization rate class, and images corresponding to the passage of the phosphate material to the rejection of the screens (phosphate loss class). The distribution of dataset images over the train and test samples is resented in Table 2.

The computation for this study was undertaken using Anaconda (version 4.10.3) with Python, employing various libraries such as OpenCV, scikit-learn, scikit-image, and TensorFlow. All experiments were performed on an Asus laptop, manufactured by ASUSTeK Computer Inc based in Taipei, Taiwan, equipped with an Intel(R) Core (TM) i5 (10th Gen) processor and 18 GB of RAM. The laptop ran at 2.50 GHz with the Windows 10 operating system. This hardware configuration provided a reliable and consistent computing environment for the execution of various computational tasks, ensuring the reproducibility and accuracy of experiments conducted throughout this study.

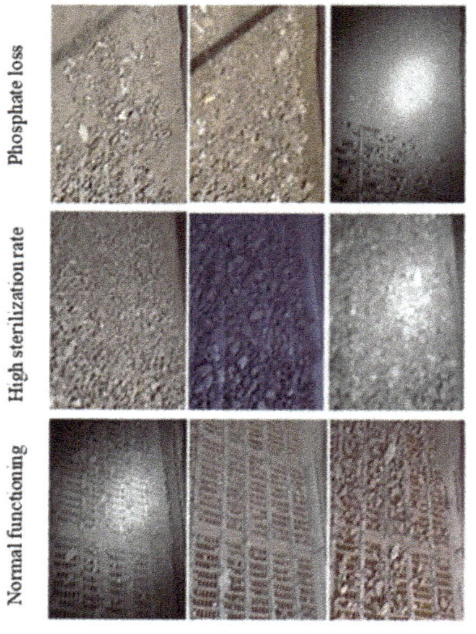

Figure 5. Dataset sample.

Table 2. Dataset distribution.

Class	Train	Test
Phosphate less	399	266
High-sterilization rate	400	267
Good functioning	401	267
Total	1200	800

3.3. Evaluation Metrics

The classification of each test sample was based on four cases commonly represented by the confusion matrix. These four cases included TP, TN, FP, and FN, corresponding to True Positive, True Negative, False Positive, and False Negative, respectively. For multi-class classification, we used a one-against-all approach as follows:

- "TP of C_i" is all C_i instances that are classified as C_i.
- "TN of C_i" is all non-C_i instances not classified as C_i.
- "FP of C_i" is all non-C_i instances that are classified as C_i.
- "FN of C_i" is all C_i instances not classified as C_i.

To compare these models' robustness, we estimated the models' accuracy, sensitivity, and specificity. Accuracy gives us an idea of the proportion of correctly classified images (TP and TN) compared to the overall number of images entered into the model (TP, TN, FP, and FN). Sensitivity is a metric that measures the model's capacity to predict each available class's True Positives. On the other hand, specificity measures the model's capacity to predict the true negatives of each available class. The equations of these metrics are as follows:

$$\text{Accuracy} = \frac{\text{TP} + \text{TN}}{\text{TP} + \text{TN} + \text{FP} + \text{FN}} \tag{1}$$

$$\text{Sensitivity} = \frac{\text{TP}}{\text{TP} + \text{FN}} \tag{2}$$

$$\text{Specificity} = \frac{\text{TN}}{\text{TN} + \text{FP}} \tag{3}$$

4. Implementation and Results

4.1. Implementation

4.1.1. Machine-Learning Approach

The machine-learning approach combines a descriptor for extracting characteristic elements from the image and a classifier forming two blocks. The evaluation is based on supervised machine learning, which consists of two phases: a training phase in which the model learns from labeled data and a test phase to assess how well the model learns from unlabeled data. The implementation process is consistent across all models in this study and involves several key stages. Initially, required libraries are imported, followed by a definition of model parameters. Specifically, the HOG technique employs nine gradient orientations, with each cell covering a 16 × 16-pixel region and two cells included in each block. SIFT utilizes a 6-pixel step between key points, effectively reducing the feature vector's size and runtime. LBP is configured with a circle radius (R) of one, circularly symmetric neighbors set points (P) equal to eight times the radius, and a uniform method to determine patterns. Our SVM used the radial basis function as its kernel with a C value of 100. For the random forest, we used 100 trees in the forest, while the KNN classifier retained its default parameters. Subsequently, the image dataset was imported, labeled, and prepared for training. The predictive model was then trained using the cross-validation method to avoid over-fitting. Finally, the model's performance was evaluated with new images from the test dataset, and various evaluation metrics were calculated. This comprehensive process ensures the appropriate development and evaluation of each model under consideration.

4.1.2. Deep-Learning Approach

The critical difference of the deep learning approach is that it combines the two stages of feature extraction and classification in a single block while exploiting the power of neural networks. This idea is based on a trainable system consisting of modules corresponding to a processing step. The training of each module is performed with adjustable parameters such as linear classifier weights. The whole system is driven from scratch: for each sample, all parameters of the modules are adjusted to match the outcome of the system to the desired outcome. The in-depth qualifier is due to the successive layering of these modules.

The architecture of the CNN model that we implemented and tested on our dataset is detailed in Table 3. This model was not pre-trained; we learned it from scratch. It included two convolution layers, each producing 64 feature maps using a 4 × 4-pixel size filter, two max-pooling windows of size 2 × 2 pixels, two batch normalization layers, two dropout layers, and three fully connected layers (FC). The final classification was achieved using the SoftMax activation function. This model follows a modeling structure comprising several vital stages. First, the necessary libraries were imported, followed by the definition of model parameters. Next, the model was created, and the images in the dataset were prepared, resized, labeled, and augmented. This model was then trained over 40 epochs. Finally, a complete evaluation was carried out, including performance tests and the calculation of key evaluation metrics.

Table 3. CNN model architecture.

Input: Image (180, 120, 1)
Normalization
Conv4-64
Maxpol-2
Dropout (0.1)
Conv4-64
Maxpool-2
Dropout (0.3)
Flatten
Fc-256
Dropout (0.5)
Fc-64
Normalization
SoftMax

Convk-m: Convolution layer with m filters whose kernel has a dimension of k × k. Maxpool-k: window pooling layer of k × k. Fc-n: multilayer n-neuron perceptron. Dropout (p): dropout with a probability of p.

4.2. Results

We evaluated several classification models; HOG and SVM, HOG and RF, HOG and KNN, SIFT and SVM, SIFT and RF, SIFT and KNN, LBP and SVM, LBP and RF, LBP and KNN, and the CNN model. The learning accuracy results presented in Table 4 reveal that most models efficiently learned data features and could provide accurate predictions during the learning phase, with learning accuracies above 80%. However, it is worth mentioning that the LBP and SVM models achieved the lowest learning accuracy of 42%. It suggests that this combination encountered difficulties in learning effectively from the training data. On the other hand, the LBP and KNN models achieved a moderate learning accuracy of 76%

Table 4. Training accuracy of different models.

Model	**HOG & SVM**	**LBP & SVM**	**SIFT & SVM**	**HOG & RF**	**LBP & RF**	**SIFT & RF**	**HOG & KNN**	**LBP & KNN**	**SIFT & KNN**	**CNN**
Train Accuracy	0.99	0.42	0.84	0.97	0.93	0.94	0.99	0.76	0.81	1

Figure 6 shows a heatmap illustrating the performance measures, including accuracy, sensitivity, and specificity, for the test dataset across different classification models explored in this study. The CNN model emerged as the best-performing model, with the highest accuracy, specificity, and sensitivity (99.6%, 99.6%, and 99.7%, respectively, as shown in Figure 6). The HOG-based models (HOG and SVM, HOG and RF, HOG and KNN) consistently performed well on all measures (exceeding 98%), especially for HOG and SVM, which achieved high accuracy (99%) and well-balanced sensitivity and specificity (99.6% and 99%, respectively). The SIFT-based models (SIFT and SVM, SIFT and RF, SIFT and KNN) also performed competitively, in particular the SIFT and RF model, which excelled in terms of accuracy and sensitivity (98% and 98.8%, respectively). By contrast, the -based models (LBP and SVM, LBP and RF, LBP and KNN) tended to provide lower accuracy and specificity. For example, the LBP and SVM models produced an accuracy and specificity of 33%, while the models based on HOG and SIFT yielded better results with higher values.

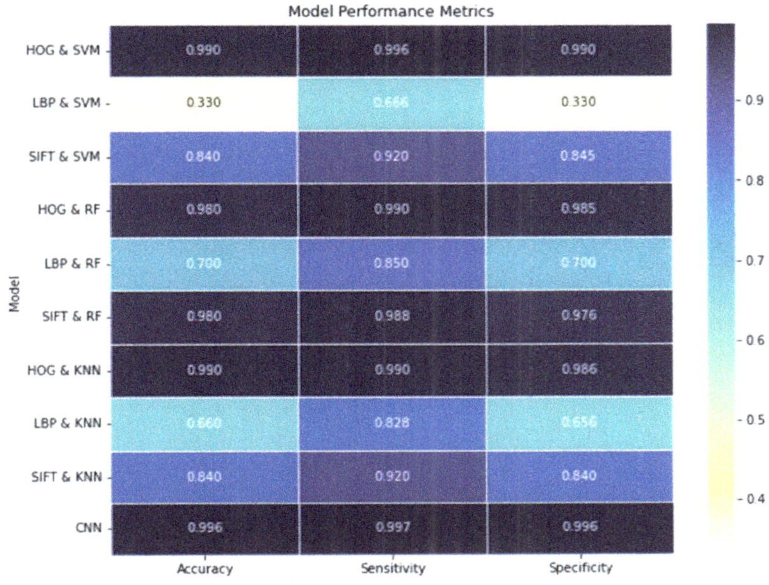

Figure 6. Heatmap of the performance metrics (accuracy, sensitivity, and specificity) for each model.

Moreover, it is worth noting that the RF classifier demonstrated consistent competence across different feature extraction methods. By contrast, the KNN classifier performed competitively despite having a slightly lower accuracy than SVM and RF.

In summary, considering the parameters evaluated, the CNN and HOG-based models were strong performers for achieving high robustness, with SIFT-based models proving competitive. Figure 7 illustrates the prediction results generated by the CNN model for a subset of images from the test sample.

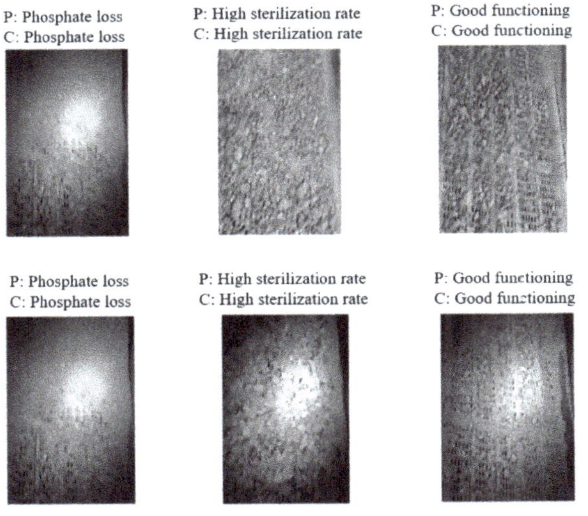

Figure 7. Prediction results of some test images using the CNN model. "P" refers to the class predicted by the CNN model, and "C" designates the actual class of the image.

5. Discussion

The previous section provided an overview of our study's findings, highlighting the performance of the CNN, HOG, and SIFT-based models in the context of anomaly detection in the screening unit of the Benguerir phosphate mine. The LBP descriptor consistently showed better sensitivity than accuracy and specificity in all combinations with various classification methods (SVM, RF, and KNN). A notable observation was made for the LBP-SVM combination. Thus, this algorithm tends to fail in describing "True Negative" instances compared to "True Positive" instances. This limitation can be attributed to the limited discriminating power of LBP, which, in some cases, may struggle to capture subtle differences between textures. This factor led us to eliminate the LBP-based models for our specific case study.

To provide a more in-depth analysis of our findings, it is crucial to discuss the real-time aspect of the intended monitoring system. In fact, the image processing time was a critical factor in our real-time system, which required the processing of two images per second with a product residence time of 10 s on the screen. An analysis of the processing times of best-performing models, as shown in Table 5, revealed that all models met the stringent processing time requirements, with each model taking less than half a second to process a single image. In particular, the SIFT-based models and the CNN model showed the highest processing speed. Consequently, if we consider both success rates and execution times, the CNN model, along with the HOG and SIFT-based models, proved to be the most appropriate choices for our case study in terms of robustness and processing speed.

Table 5. Processing time of an image for the models with highest accuracies.

Model	Time to Process an Image (s)
HOG and SVM	0.025
HOG and RF	0.013
HOG and KNN	0.013
SIFT and SVM	0.004
SIFT and RF	0.0005
SIFT and KNN	0.0007
CNN	0.0008

Furthermore, comparing the classical approach with the deep approach for convolution neural networks, the advantage of the deep architecture is that it is not necessary to build a feature extractor by hand since all these layers are trained to extract features in the image in an automatic way. In ref. [13], we evaluated and compared the performances of the CNN as a descriptor and the HOG, SIFT, and LBP descriptors, each using an SVM classifier. As a result, we found that the deep neural network approach is robust and offers the greatest accuracy despite a low runtime trade-off.

6. Conclusions

Intelligent monitoring systems require the use of powerful and robust models that are capable of accurately fulfilling their assigned purpose. In this paper, we present a case study in which we provide a comparative study of different classification models designed to accurately detect anomalies in the screening unit of the Benguerir phosphate mine.

The experimentation in this research section highlights the robustness of both the CNN and HOG-based models. The CNN model demonstrated exceptional accuracy, specificity, and sensitivity, all above 0.99. Simultaneously, the HOG-based models performed well, with accuracy, specificity, and sensitivity all exceeding 0.98. Notably, both models achieved these results while maintaining a highly tolerant processing speed. While the SIFT-based models did not match the performance of the CNN- and HOG-based models, they still achieved competitive results. These outcomes were obtained based on a dataset of images taken under normal conditions. However, the mine is an uncertain environment subject to severe weather conditions (fog, dust, rain, and high temperature). Hence, precise knowledge of these methods' robustness in images containing parasites and noise caused by degraded weather conditions or other noise sources is imperative. Indeed, this concern is the major challenge of any artificial vision system in a context like mining.

Our perspective for the next step is to examine the noises and degradations that can alter the quality of images captured from surveillance cameras and then to study suitable solutions to rectify these defects. After this system is complete, the overall objective is to integrate this system with other systems, such as serving autonomy, to develop a global digital platform that ensures the remote control of all activities of the mine in Benguerir.

Author Contributions: Investigation, L.E.H.; Methodology, L.E.H.; Software, L.E.H.; Supervision, A.E. and N.A.; Validation, A.E. and N.A.; Writing—original draft, L.E.H.; Writing—review and editing, A.E. and N.A. All authors have read and agreed to the published version of the manuscript.

Funding: The UM6P and the National Centre for Scientific and Technical Research provided funding for this work.

Data Availability Statement: Data is unavailable due to privacy issues.

Conflicts of Interest: The authors declare no conflict of interest.

References

1. Kelly, T.; Matos, G.R.; Buckingham, D.A.; DiFrancesco, C.A.; Porter, K.E.; Berry, C.; Crane, M.; Goonan, T.; Sznopek, J. *Historical Statistics for Mineral and Material Commodities in the United States*; Data Series; U.S. Geological Survey: Reston, VA, USA, 2005; Volume 140.
2. Steen, I. Phosphorus Availability in the 21st Century: Management of a Non-Renewable Resource. *Phosphorus Potassium* **1998**, *217*, 25–31.
3. Cordell, D.; Drangert, J.-O.; White, S. The story of phosphorus: Global food security and food for thought. *Glob. Environ. Chang.* **2009**, *19*, 292–305. [CrossRef]
4. Van Vuuren, D.P.; Bouwman, A.F.; Beusen, A.H.W. Phosphorus demand for the 1970–2100 period: A scenario analysis of resource depletion. *Glob. Environ. Chang.* **2010**, *20*, 428–439. [CrossRef]
5. Stewart, W.M.; Dibb, D.W.; Johnston, A.E.; Smyth, T.J. The Contribution of Commercial Fertilizer Nutrients to Food Production. *Agron. J.* **2005**, *97*, 1–6. [CrossRef]
6. Gouaillier, V.; Fleurant, A.-E. *La Vidéosurveillance Intelligente: Promesses et Défis*; CRIM: Montreal, QC, Canada, 2009; Volume 104.
7. David Jenkins, M.; Buggy, T.; Morison, G. An imaging system for visual inspection and structural condition monitoring of railway tunnels. In Proceedings of the 2017 IEEE Workshop on Environmental, Energy, and Structural Monitoring Systems (EESMS), Milan, Italy, 24–25 July 2017; pp. 1–6.
8. Marino, F.; Distante, A.; Mazzeo, P.L.; Stella, E. A Real-Time Visual Inspection System for Railway Maintenance: Automatic Hexagonal-Headed Bolts Detection. *IEEE Trans. Syst. Man Cybern. Part C* **2007**, *37*, 418–428. [CrossRef]
9. Park, S.; Choi, Y. Applications of Unmanned Aerial Vehicles in Mining from Exploration to Reclamation: A Review. *Minerals* **2020**, *10*, 663. [CrossRef]
10. Dey, P.; Chaulya, S.K.; Kumar, S. Hybrid CNN-LSTM and IoT-based coal mine hazards monitoring and prediction system. *Process Saf. Environ. Prot.* **2021**, *152*, 249–263. [CrossRef]
11. Wang, C.; Chang, L.; Zhao, L.; Niu, R. Automatic Identification and Dynamic Monitoring of Open-Pit Mines Based on Improved Mask R-CNN and Transfer Learning. *Remote Sens.* **2020**, *12*, 3474. [CrossRef]
12. El Hiouile, L.; Errami, A.; Azami, N.; Majdoul, R.; Deshayes, L. Evaluation of Classical Descriptors coupled to Support Vector Machine Classifier for Phosphate ore Screening monitoring. In *Proceedings of the 13th International Conference on Intelligent Systems: Theories and Applications, Rabat, Morocco, 23–24 September 2020*; Association for Computing Machinery: New York, NY, USA, 2020; pp. 1–7.
13. El Hiouile, L.; Errami, A.; Azami, N.; Majdoul, R. Deep Neural Network Descriptor for Anomaly Detection in the Screening Unit of an Open Pit Phosphate Mine. In *International Symposium on Ubiquitous Networking*; Elbiaze, H., Sabir, E., Falcone, F., Sadik, M., Lasaulce, S., Ben Othman, J., Eds.; Springer International Publishing: Cham, Switzerland, 2021; pp. 239–251.
14. Laaniber, N.; El Krouni, M. Rapport du Projet de fin d Études—PDF Téléchargement Gratuit. Available online: https://docplayer.fr/82250671-Rapport-du-projet-de-fin-d-etudes.html (accessed on 20 May 2023).
15. Tidriri, K. Decision Fusion Dedicated to the Monitoring of Complex Systems. Ph.D. Thesis, Université d'Angers, Angers, France, 2018.
16. Tidriri, K.; Chatti, N.; Verron, S.; Tiplica, T. Bridging data-driven and model-based approaches for process fault diagnosis and health monitoring: A review of researches and future challenges. *Annu. Rev. Control.* **2016**, *42*, 63–81. [CrossRef]

17. Weimer, D.; Scholz-Reiter, B.; Shpitalni, M. Design of deep convolutional neural network architectures for automated feature extraction in industrial inspection. *CIRP Ann.* **2016**, *65*, 417–420. [CrossRef]
18. Staar, B.; Lütjen, M.; Freitag, M. Anomaly detection with convolutional neural networks for industrial surface inspection. *Procedia CIRP* **2019**, *79*, 484–489. [CrossRef]
19. Haselmann, M.; Gruber, D.P.; Tabatabai, P. Anomaly Detection Using Deep Learning based Image Completion. *arXiv* **2018**, arXiv:1811.06861.
20. Ricco, R. SVM—Support Vector Machine. Available online: https://eric.univ-lyon2.fr/ricco/cours/slides/svm.pdf (accessed on 1 October 2023).
21. Pal, M. Random forest classifier for remote sensing classification. *Int. J. Remote Sens.* **2005**, *26*, 217–222. [CrossRef]
22. Cunningham, P.; Delany, S. k-Nearest neighbour classifiers. *arXiv* **2007**, arXiv:2004.04523.
23. Khan, S.; Rahmani, H.; Shah, S.A.A.; Bennamoun, M. A Guide to Convolutional Neural Networks for Computer Vision. In *Lectures on Computer Vision*; Springer: Cham, Switzerland, 2018; Volume 8, pp. 1–207. [CrossRef]
24. Seemanthini, K.; Manjunath, S.S. Human Detection and Tracking using HOG for Action Recognition. *Procedia Comput. Sci.* **2018**, *132*, 1317–1326. [CrossRef]
25. Patel, H.A.; Rajput, R.D. Smart Surveillance System Using Histogram of Oriented Gradients (HOG) Algorithm and Haar Cascade Algorithm. In Proceedings of the 2018 Fourth International Conference on Computing Communication Control and Automation (ICCUBEA), Pune, India, 16–18 August 2018; pp. 1–4.
26. Vashistha, P.; Bhatnagar, C.; Khan, M.A. An architecture to identify violence in video surveillance system using ViF and LBP. In Proceedings of the 2018 4th International Conference on Recent Advances in Information Technology (RAIT), Dhanbad, India, 15–17 March 2018; pp. 1–6.
27. Bilik, S.; Horak, K. SIFT and SURF based feature extraction for the anomaly detection 2022. *arXiv* **2022**. [CrossRef]
28. Amraee, S.; Vafaei, A.; Jamshidi, K.; Adibi, P. Abnormal event detection in crowded scenes using one-class SVM. *SIViP* **2018**, *12*, 1115–1123. [CrossRef]
29. Decor, G.; Bah, M.D.; Foucher, P.; Charbonnier, P.; Heitz, F. Defect Detection in Tunnel Images using Random Forests and Deep Learning. In Proceedings of the 10th International Conference on Pattern Recognition Systems, Tours, France, 8–10 July 2019.
30. Dalal, N.; Triggs, B. Histograms of Oriented Gradients for Human Detection. In Proceedings of the 2005 IEEE Computer Society Conference on Computer Vision and Pattern Recognition (CVPR'05), San Diego, CA, USA, 20–25 June 2005; IEEE: San Diego, CA, USA, 2005; Volume 1, pp. 886–893.
31. Lowe, D.G. Distinctive Image Features from Scale-Invariant Keypoints. *Int. J. Comput. Vis.* **2004**, *60*, 91–110. [CrossRef]
32. Ojala, T.; Pietikäinen, M.; Harwood, D. A Comparative Study of Texture Measures with Classification Based on Featured Distributions. *Pattern Recognit.* **1996**, *29*, 51–59. [CrossRef]

MDPI
St. Alban-Anlage 66
4052 Basel
Switzerland
www.mdpi.com

Mining Editorial Office
E-mail: mining@mdpi.com
www.mdpi.com/journal/mining

www.ingramcontent.com/pod-product-compliance
Lightning Source LLC
Chambersburg PA
CBHW061514210926
51850CB00008B/1916
* 9 7 8 3 0 3 6 5 9 9 4 1 0 *